The Brown Bess

An Identification Guide and Illustrated Study of Britain's Most Famous Musket

by Erik Goldstein & Stuart C. Mowbray

Curator of Mechanical Arts and Numismatics,
Colonial Williamsburg

Editor, *Man at Arms for the Gun and Sword Collector* magazine

Mowbray Publishing • 54 East School Street • Woonsocket, RI 02895 USA • www.gunandswordcollector.com

Erik Goldstein & Stuart C. Mowbray
The Brown Bess: An Identification Guide and Illustrated Study of Britain's Most Famous Musket
54 East School St., Woonsocket, R.I. 02895: ANDREW MOWBRAY INC. — PUBLISHERS
160 pp.

ISBN: 1-931464-44-8

© 2010 Erik Goldstein and Stuart C. Mowbray

All rights reserved. No part of this book may be reproduced in any form or by any means without permission in writing from the publisher, except for short excerpts used in book reviews.

Printed in China.   9 8 7 6 5 4 3 2 1

To order more copies of this book or to carry it in your store, museum or gunshop, call 1-800-999-4697 or email: orders@manatarmsbooks.com. We carry a full line of fine gun and sword collecting books; a free catalog is available upon request.

Note about the illustrations: Unless credited otherwise, all photographs are by Stuart C. Mowbray and cannot be scanned or otherwise copied without permission. Many of the muskets appearing in this book are from the collections of Colonial Williamsburg and can be visited at their Dewitt Wallace Museum.

No book of this size and scope could be written without help and inspiration. We especially want to tip our hat to the incomparable De Witt Bailey, our friend and mentor who has done more to advance the study of these muskets than any other researcher. His books on British military weaponry are "must have" resources for any serious student of flintlock muskets, and we can honestly say that this book would never have been written without his ground-breaking work. We would also like to thank Don Troiani, who let us paw through his collection of muskets and photograph a couple of them for inclusion here. Joseph V. Puleo was very helpful with research assistance and proofreading. George C. Neumann allowed us to photograph two of his muskets and was equally generous with his extensive knowledge. Jim Gooding is to be thanked for his wise guidance and information on obscure London gunmakers. Peter Schmidt and R. Scott Stephenson helped with proofreading. Chris Fox from Fort Ticonderoga saved our bacon on more than one occasion and we still owe him a beer. One of our photographs was also taken at their annual Revolutionary War reenactment, an enjoyable event that we highly recommend. Billy Ahearn pitched in with proofreading and let us study and photograph his collection. Lastly, we need to acknowledge the late Anthony Darling for encouraging both of us in our youth, as he did for so many other young collectors of our generation. And if Chuck was here today, he would certainly remind us to credit the late Howard Blackmore for his ground-breaking research, which continues to set the standard for excellence — and probably always will.

This book is dedicated to our wives, Chelsea (a.k.a. "the shiksa") and Kim. Our friends insist that we do not deserve such smokin' hot, yet shockingly intelligent, women and doubtless they are right.

TABLE OF CONTENTS

Foreword	4
Brown Bess Basics	5
Markings Commonly Found on Brown Bess Muskets	14
A Very Brief Look at Bayonets	16
A Quick Identification Guide	17
The Pattern 1730 Long Land Musket	18
The Pattern 1730/40 Long Land Musket	30
The Pattern 1742 Long Land Musket	38
The Pattern 1748 Long Land Musket	50
The Pattern 1756 Long Land Musket (British)	62
The Pattern 1756 Long Land Musket (Irish)	74
The Militia or Marine Muskets (Patterns 1757 & 1759)	84
The Pattern 1769 Short Land Musket (British)	92
The Pattern 1769 Short Land Musket (Irish)	102
The Pattern 1777 Short Land Musket	112
The Liège Short Land Pattern Muskets	122
The Pattern 1779-S Short Land Musket	132
The Pattern 1793 Musket (India Pattern Type 1)	142
The Pattern 1809 Musket (India Pattern Type 2)	152
Bibliography and Suggestions for Further Reading	160

The ruins at Crown Point.

FOREWORD

Be they factual or fictional, legendary warriors throughout history have always carried equally legendary and iconic weapons. King Arthur would have been nothing without his Excalibur. Fearless Zulu Warriors wielded their distinctive assegais when they charged at Rorke's Drift. It was a Smith & Wesson Model 29 Revolver that "Dirty Harry" used to perforate hapless villains, whether they felt lucky or not. And for the famed British Red Coats, that legendary weapon was the Brown Bess Musket. The exact roots of the term "Brown Bess" are unknown, lost in the melange of 18th-century military lore, but the nickname is obviously an affectionate one. Bess was a sturdy, dependable wench who felt good in your hands during desperate moments.

This book is organized chronologically by Pattern date. Each basic Pattern of musket will be illustrated by one or more examples representing the archetypal forms of that Pattern as it was assembled and issued. Few flintlock muskets survive today with all of their original parts, but we have done our best to show as many "as manufactured" features as possible. The Pattern date system that we have employed is the creation of author and researcher De Witt Bailey. It is a precise and convenient way for modern collectors and scholars to identify individual muskets and then clearly communicate what they are looking at. However, it should be stressed that these Pattern dates were never official military or Ordnance terms.

During the years when these muskets were being carried in service, Ordnance workers tended to use terms like "Old King's Pattern" or "New King's Pattern" to tell musket types apart – they rarely got any more specific than that. And of course, over time, the "New King's Pattern" would become the "Old King's Pattern" as further design changes were adopted. But for soldiers in the field, all of these muskets would have looked the same. Whether it was a so-called "Pattern 1730" or a so-called "Pattern 1756," it was just a musket. It is easy for modern collectors (who just love to pigeonhole everything) to forget that 18th and early 19th-century soldiers never saw any truly identical objects – at least not things that we would consider identical today. Just think of how many identical things we see in our daily lives. From Coke bottles to dishwashers, most modern manufactured goods are perfect clones with millions upon millions of absolutely identical brothers and sisters scattered across the globe. But Tommy Red Coat never saw a Coke bottle. For the most part, he only saw handcrafted goods with no interchangeable parts. So whether his musket was the new pattern or the old one, it was still probably the single most identical thing that he had ever seen in his entire life. The small differences that thrill modern collectors would have meant nothing to him. To him, they were all the same.

It was only with the introduction of what we call the "Short Land Pattern of 1769" that the basic "Brown Bess" musket changed enough for it to become noticeable. This new musket was a lot shorter than the old one, so we see Ordnance references to "Short Land Pattern" as opposed to the old standby "Long Land Pattern." But even this major change wouldn't have meant much to Red Coats in their various regiments. Sure, the new short muskets were handy (and as such were given first to the Light Infantry), but they still looked and worked pretty much the same. It certainly wasn't a big enough difference to warrant removing the long pattern muskets from service, and the short and long patterns were often carried right along side of each other.

Now, of course, we are going to spend about a hundred and sixty pages obsessing about those little, unimportant differences that Tommy Red Coat never noticed. We are, after all, historians...not soldiers.

Courtesy of the Anne S.K. Brown Military Collection at Brown University.

BROWN BESS BASICS

We hate to say it, but most historians and collectors are totally clueless when it comes to Brown Bess muskets. But it isn't their fault. This can be complicated stuff, and any newcomer to the topic is sure to be intimidated by the sheer number of musket variations that exist and the seemingly invisible details that can tell them apart.

The muskets, themselves, don't help very much, either. Plenty of Brown Besses that you see today are actually cobbled together pieces of junk made up from parts that don't match. Some of these guns were assembled in historical times, but just as many were put together more recently by unscrupulous jerks looking to make a quick buck. Anyone trying to identify one of these Frankenstein's Monsters using the available reference books is sure to end up frustrated. As it stands, the basic identification of even the most common Brown Bess types stumps most dealers, museum curators and auction houses, leading to a situation where there are just a handful of guys who everyone keeps calling when they are faced with a musket that has different or interesting features. Some of these gurus know what they are talking about…others don't.

However, there is no reason for this situation to continue. In fact, most Brown Besses are amazingly easy to understand. All you need is a little information at your fingertips. But where can you locate this kind of data? Even in the best reference books and magazine articles, good photos and descriptions of the essential identifying features that you will need to recognize in order to tell the basic models apart are almost impossible to find. In fact, quite a few of these publications are only useful if you are *already an expert* on the topic.

That's where this book comes in. We have taken what can be a very confusing topic and we have boiled it down to its essentials – all illustrated with a vast assortment of huge, clear, color photos showing all the important details that you are looking for. And we have kept things simple. Our organization is strictly chronological and each musket type has its own chapter – with a full ten or more pages of photos showing every feature in remarkable detail – so there is just no way to get lost or confused. If you've got a Brown Bess in your lap right now, and can't wait to find out what you have, then by all means, skip straight to page 17 where we have provided a quick identification guide. That way, you can look your musket up in the Table of Contents and move on to the chapter dedicated to that specific weapon. But for everyone else, let's start with some "Brown Bess Basics."

First of all, while American collectors are accustomed to calling musket types "models," this is not appropriate for British military firearms – we use the word "pattern" instead. Each musket pattern has a date assigned to it, indicating the year in which it was introduced. Every one of these patterns will have its own chapter in this book. Pretty simple, huh?

As was mentioned in our Foreword, the pattern date system that we have chosen is based loosely upon the research publications of author De Witt Bailey – all of which are highly recommended to our readership. However, we must make clear that ours is a highly simplified and somewhat modified version of his approach. Bailey's system is drawn from Ordnance Department records and every minor change to a design, or arsenal modification to an existing musket, receives a separate, new pattern designation. The resulting system is amazingly complete and is a powerful tool in the right hands, but it has proven to be a bit unwieldy for average readers who simply want to identify the basic musket types. For example, some of Bailey's "patterns" are entirely hypothetical – meaning that their features are mentioned in a piece of historical correspondence, but no matching muskets have been observed to confirm that these items were ever actually made. Since this is meant to be a useful, hands-on type of book, rather than a purely academic study, we hesitate to include muskets that might only exist in theory.

With this in mind, our pattern dates are based purely upon cases where significant numbers of surviving muskets are available to confirm what is recorded in the paperwork. We have also boiled the system down further, so that simple repair work or upgrades (like swapping a steel ramrod for a wooden one) do not necessitate a change in the musket's designation. The same thing goes for truly exotic musket variations that hardly anyone ever sees. Our intention is to create a practical guide to basic musket types; we will leave the one-of-a-kind rarities to other authors and concentrate upon the 99.9% of muskets that you are actually likely to encounter. We feel that the resulting system does honor to Bailey's ground-breaking research while, at the same time, making it much easier to use for the vast majority of collectors, curators and historians.

In order to understand Brown Bess muskets, you need to understand how they were made and how they were marked. All Brown Besses procured by the British Ordnance are covered with proof marks, maker's marks, assembly numbers, property marks and storekeeper's marks. These markings are the code that might allow you to decipher whether a Brown Bess has all of its original parts. They can also tell you quite a bit about a musket's history.

All of these markings are evidence of something we call the Ordnance System of Manufacture. The British Board of Ordnance, which oversaw the purchasing and issue of weapons for the British military, had a strict system for manufacturing and organizing their small arms. First, starting as early as the late 17th century, muskets were defined as being either for "Sea Service" or for "Land Service." So when we say that a musket is a "Long Land Pattern," we mean that it is a land service musket with a lengthy barrel. A "Short Land Pattern" is a similar gun with a shorter barrel. Get it? Then, about 1714, the Ordnance Department decided to do something *really* radical, aimed at purchasing higher quality muskets, all fully inspected and made to very specific patterns. In other words, they wanted to get a better, more consistent product. And while they knew that the parts of these muskets would never be perfectly interchangeable, they wanted to do their very best at making everything uniform in order to simplify repairs in the field. The way that they decided to accomplish this was to avoid buying whole, completely finished muskets from gunmakers. That's because no matter how much they tried, each gunmaker would make his own muskets a little different. The only way for the Ordnance to achieve the uniformity that it was looking for was to buy all the parts separately, from specialized workmen, and then have these nearly identical pieces assembled into complete muskets at a later time. In fact, they often stockpiled parts for years before using them for musket assembly. Another benefit of this system was that all these stockpiled parts made it really easy for the Ordnance Department to quickly assemble piles of extra muskets when a war broke out and demand suddenly increased. It was, without exaggeration, a stroke of genius. As will be explained in the specific chapters of this book, while this Ordnance System of Manufacture was modified in very small ways over the years, it remained the cornerstone of British musket procurement for nearly a century.

This is how it worked. First, a pattern would be developed and approved for each and every component of a Land Service musket. Then the Board of Ordnance, from its headquarters at the Tower of London, would make contracts with individual craftsmen, often located in the nearby Minories neighborhood. When they were done, these musket lock makers, barrel makers, rough stockers, bayonet makers, furniture makers, etc. would submit their products to the Tower for inspection. Locks were the most critical element of any musket, and we are lucky enough to have a very specific description of what happened next. On May 21, 1760, there was a robbery trial in London, and Thomas Hartwell, who worked "under the surveyer general in the small gun office" of the Ordnance Department, testified about how locks were inspected:

...locks are sent in by the contractors in this state (holding a soft lock in his hand) with the contractor's mark on them. (He takes up a hard one) this is mark'd with all his Majesty's proper marks... Before the King's mark is put upon them...they are narrowly inspected by a proper officer, to see whether they are according to the pattern agreed for; if they are approv'd of, he immediately strikes a proper mark on the inside, which we all know, and all the trade knows; if we meet with a lock in the East Indies we know who it was viewed by...he strikes his mark first, and immediately after that strikes the broad arrow on the outside... then they go to the engravers, then to the hardeners; they go through another inspection after that, and if needful they are return'd to be made good.

From Hartwell's testimony, we learn that locks were viewed, accepted, stamped with an inspector's mark on the inside and with a government ownership mark on the outside, disassembled, engraved, hardened, reassembled and then inspected again.

The inspector's mark on the inside of a lockplate was a crowned numeral assigned to a particular inspector...much like the "Inspected by Number 14" paper tag that you sometimes find floating around when you open a package of new underwear. It should be noted that just because a contractor was hired to supply a particular set of parts, that does not necessarily mean that he did the work himself. There were many subcontracts, and sometimes the initials of subcontractors can be seen on the insides of lockplates, etc.

The crowned broad arrow that we find struck into the exterior face of Ordnance musket lockplates was called the "King's mark." It identified any item thus marked as being crown property. The mark was well known and people would literally stop you in the street if they saw you carrying something marked in that way. Any serious study of Georgian newspapers or courtroom testimony will reveal repeated instances when the King's mark was used to identify items as stolen from the government. The mark, in and of itself, was enough to get you convicted. No further proof was required. In a humorous aside to all this, one courtroom record describes how a nervous drunkard illicitly put the broad arrow on his bag of cocoa nuts so that other people would be too scared to steal them!

Once accepted, all of these locks, ramrods and other parts might be stored away in barrels for future use or they might be used up immediately, depending upon the situation. Under the Ordnance System, muskets were not generally assembled at the Tower, but by outside gunmakers who contracted to perform the "rough stock & set up." Thomas Ashton, an Ordnance contractor, described the system this way:

I am a gun-maker, I work for the Tower; I receive out barrels and locks from the Tower...they are all new [and] they have the King's mark upon them... I put them on stocks, and make them into guns, and return them into the Tower.

In April of 1812, he was doing this work with the help of two journeymen, Joseph Bull, who had worked for him two years, and Thomas Kipping, who had been on the job four months. (Proceedings of the Old Bailey, reference number t18120408-72).

All of the parts being used to set up a particular musket would usually be marked with an assembly number in the form of a roman numeral. This allowed disassembled parts belonging to different muskets to be differentiated in the workplace. Since locks had so many parts of their own, lockmakers often used the same system, so there are often two sets of assembly numbers on any given musket, one set for the overall weapon and another for the individual parts of its lock.

When this setting up was finished, the assembled muskets would be returned to the Tower where they would be inspected and assigned for storage or issue. This was when one or more "storekeeper's marks" would be struck into the wooden stock, indicating that the musket had been taken into store. These many markings did more than ensure quality and provide some accountability; they were also a critical part of

It's Not Easy Being a "Chuckle-headed Dog"!

On September 7th, 1737, Francis Fuller of St. Margaret's, Westminster, Esq., 1st Major of the 1st Foot Guards with a double rank of Lt. Colonel, "not having God before his eyes, but being moved and seduced by the Devil," was charged with murdering a soldier under his command. Witnesses who were in St. James' Park, where the regiment drilled twice per day, described this scene:

Colonel Fuller [was] exercising the Sergeants and Corporals on Friday morning. The Colonel sent all the awkward ones, – those that Exercised awkwardly, to a plain place in the park [called] the Hole, to the Drill Serjeants. The Moment they don't Exercise right, they are sent down to Drill-Serjeant Joseph Gage, for further Exercise, – to learn further. After the Men were dismiss'd, the Colonel went down to the Drill-Serjeant's to see [the awkward ones] exercise himself.

[Colonel Fuller] walked along a single Rank, and found fault with some People, before he came to Adam Cluff, a soldier in Holbert's Company. But when he came to him, he call'd him a Chuckle-headed Dog for not Exercising Right, [saying] that he kept his Elbow too high. [Cluff] did not take it lower. That put the Colonel in a Passion with him. He struck him with his Fist under the right and left side of his Jaws and his Throat. [Cluff] had a Firelock upon his Shoulder at the time [of the] three blows.

When he performed that Part of His Exercise, which they call resting the Musket, the Colonel catch'd the firelock with his Hands, and jabb'd it against his left Breast, because it did not lye right under his Shoulder. He put it upon [Cluff's] Shoulder, telling him at the same Time that he did not do the Motion that he had shewn him before. He took it out of his Hands and jabb'd the Guard of the Firelock against his Breast. [Not the] Butend, but the Guard of the firelock, that is the Part where the Trigger is. He made him rest his Firelock according to this new Method of theirs, – with the left Hand across the Breast. The Colonel did it to make him hold it lower; he did not rest his Piece as he should have done, so the Colonel took it out of his Hands, shewed him how to rest it with a proper Motion, then he went at him, and jabb'd this sharp part against his Breast, and threw the Stock across his Shoulder, telling him if he did not do better in the Afternoon than he had done in the Morning, he would send him to the Savoy. Then he went away.

On the Wednesday Evening following, about 7 o'clock, at a Gin Shop in Chapel Street, Westminster, Cluff came by the door, stooping, bending and coughing. "Oh! says he, (with his Hand across his Breast) I am a dead Man; for the Blows that Colonel Fuller gave me has occasioned my Death. My officer is not in Town, but he has sent me word that he will see me righted." On Saturday, Cluff died.

— Compiled from testimony at London's "Old Bailey" criminal court. Cluff was treated by the regimental surgeon, who bled him and gave him an oily medicine for his cough. When Cluff died, the surgeon attributed this to his "fizzy blood" and not his wounds. Fuller was charged with murder and a coroner's inquiry dug up Cluff's body. The coroner broke all of the cadaver's ribs. Since each one of them broke with a nice snapping noise, he concluded that Cluff could not have died from his beating because his ribcage was intact. They ruled "natural causes." Various non-commissioned officers testified that Fuller had only hit Cluff in the head and no more violently than was normal. Fuller was acquitted. He would rise to the rank of Major General, commanding the 29th Regiment of Foot. He died while commanding that Regiment at the Fortress of Louisburg in 1748.

Life as an Ordnance worker at the Tower was not easy. Hours were long, and if you got hurt – tough luck – you were out of a job. Records from London's "Old Bailey" criminal court show the desperation some workers were driven to after being hurt on the job. Richard Dorrell, a Tower workman, was arrested on June 20, 1780, after a local distiller witnessed him walking down Cheapside with three bayonets in his hands. The distiller noticed that the bayonets bore Ordnance marks. Dorrell was "in drink" and claimed that the bayonets were his. William Wornum, Inspector of Bayonets at the Tower, testified that the bayonets *"are the King's; there is the mark of the viewer and the mark of the contractor upon them... I am almost positive they were taken out of the shop where I work... in the Tower. I saw the prisoner about six o'clock on the Tuesday evening, the day he was* [arrested], *very much in liquor in the Tower."* When asked the value of the bayonets, Wornum responded, *"pretty near two shillings, rather under that; they are damaged and not fit for the King's service."* He also mentioned that they were unfinished. Dorrell's defense was, *"I had a hurt in my toe and was discharged, being unable to work. I took* [the unfinished, rejected bayonets] *to make me a few tools of; I did not think there was any harm in taking them."* His job at the Tower had been fitting the bayonets to the muskets. He was found guilty.

It should also be noted that all through this time, repair and refurbishing work was also being performed by Ordnance Department personnel at the Tower. Having so many muskets and other weapons being disassembled sometimes led to thievery. Brass, in particular, was easy to turn into quick money, making it a particular temptation. The Tower never sold old scraps of brass but always saved them up to be melted down and an elaborate bookkeeping system that let the Ordnance know whether a particular part or weapon was ready for issue, and whether it had been paid for yet. All of these specific marks will be explained in a guide that appears on page 14.

It should be noted that the Tower had workshops of its own, which often performed parts of the assembly and finishing work, such as polishing. Little is known about this aspect of the Ordnance's operations, but they had many workshops along the edge of the Tower complex and employed a large number of men.

Courtesy of the Anne S.K. Brown Military Collection at Brown University.

(above) The difference between a Long and Short Land Pattern barrel is exemplified by this 4-inch section hacked from the end of a barrel and discarded in a Champlain Valley, NY camp c.1777.

(below) The sheet-brass tips of wooden ramrods were held in place by a wedge driven into a cut in the end of the ramrod. This clever method of attachment was obscured by the tip, but is easily seen in this excavated example from the Champlain Valley, NY.

This is a Pattern 1756 Long Land Pattern musket issued to the 21st Regiment of Foot, i.e. the Royal North British Fuziliers, before they embarked for Canada. They surrendered *en masse* in 1777 after Saratoga (see historical sign below marking where they grounded their arms).

Although well worn and pitted, this musket is still extremely desirable.

According to standard Ordnance Department practice, when a musket was regimentally marked at the Tower, the regimental name, number or crest was engraved on the barrel top near the breech.

made into new parts. In June of 1826, a 48-year-old Ordnance Department employee named Thomas Wiseman was arrested for stealing 42 lbs. of brass sword furniture, 14 lbs. of brass pistol furniture, 7 pistols, 11 lbs. of assorted brass, 19 musket and pistol locks, 28 sword blades, 3 lbs. of iron buckles, 7 powder horns, 2 lbs. of brass wire, 2 lbs. of locks and springs, 19 steel ramrods, 6 springs and 29 wooden grips – all the property of the King. Jonathan Bellis, master furbisher of the Tower, testified that Wiseman *"has been twenty-two years in the service of the Ordnance department – his employ was to brush arms – he worked from seven in the morning till eight in the evening; he had access to the armoury. [The items he stole] are the property of his Majesty. They were returned from the various volunteer corps."* (Proceedings of the Old Bailey, 22nd June 1826, case number 1285.)

Some of the sexiest Brown Besses are those marked to specific regiments of the army. The more historical the regiment, the more intriguing the musket. Most commonly encountered are the muskets that belonged to regiments that surrendered as a group. In New England during the 1960s, it was still common to come across Brown Bess muskets marked to regiments who laid down their arms after the Battle of Saratoga. Since these British regimental markings often confuse collectors, and because they are often a key factor in determining a particular musket's value, we will give this subject some special attention here.

The idea of applying ownership markings to military firearms goes back into the early 17th century, so no one should be surprised that many, if not most, of the earliest Long Land muskets are "fully regimentally marked," meaning that these weapons carry three-part designations including:

1) The number or name of the regiment. All regiments were numbered after 1751, but many also had titles, like the 3rd Regiment, called "The Buffs," or the 18th Regiment, also known as "The Royal Irish Regt," etc. Since most examples of the first two musket patterns covered in this book were issued before the official numbering of the regiments of the British Army in 1751, the name (and often the rank) of the commanding Colonel is frequently used as the

ON THESE FIELDS
THE BRITISH ARMY
GROUNDED ARMS
AT THE SURRENDER

N.Y. STATE
HISTORICAL
MARKER
1927

9

unit's designator, and is most often engraved along the top of the barrel near the breech.

2) The Company. In the majority of cases, each battalion of the regiment was divided into 8 to 10 companies of between 38 and 100 privates, corporals and serjeants. Two of these were the elite "flank companies," the feared Grenadier and Light Infantry companies (the latter being created in 1771). However, there were units that had bigger and more companies, for a variety of reasons. The large fluctuations in numbers of companies and their size were directly related to the need for troop strength during wartime as opposed to peacetime.

3) The individual soldier/weapon number. While this needs no further explanation, it should be pointed out that officers were not included in tallies for weapons issue, as they supplied their own arms. The Board of Ordnance only supplied firearms and bayonets to privates, corporals and serjeants, the latter of which often carried carbines or fusils (instead of regular muskets) in place of a halberd.

Before the middle of the 18th century, most muskets were marked after they left Ordnance stores. This was done at the regimental Colonel's whim and expense, and marking styles vary greatly. While the regimental title or name most often appears on the barrel, these arms tend to be heavily marked with company commander's names engraved on the tang of the buttplate, their initials branded into the stocks, as well as the more usual company and weapon numbers (often separated by a line and forming the shape of a fraction) on the wristplate.

As the Ordnance started to impose an ever-increasing amount of standardization upon their muskets, they also sought to regularize the manner in which these arms were regimentally marked. So what we see developing in the years before the American Revolution is a semi-official "Ordnance style" of regimental markings. These markings followed the three-part hierarchy described earlier (regiment, company and weapon number). The Ordnance's easily recognizable format had either the name or the number of the musket's regiment followed by "REGT" (and battalion number, if they had more than one) engraved in block letters along the top of the barrel. The wristplate bore a fraction-style (i.e., "rack") marking indicating the company and weapon number of the musket. On rare occasions, the Ordnance would engrave the crest of the regi-

A.S.K. Brown collection.

Typical examples of the classic British Ordnance style of regimental marking. The barrel is marked with the name of the regiment itself. In this case, it is the 1st Battalion of the 71st Regiment, a unit that only existed during the Revolutionary War. This wristplate from another musket has a "fraction"-style "17" over "73", meaning it is the 73rd musket in the 17th Company.

(above) **Issued from Dublin Castle, the arms of the 44th Regiment were distinctively marked. These marking were applied by the regiment. This is musket number 6 of Company K and the bayonet of musket 15 of Company E.**

(below) **These two wrist plates, found in the Caribbean, illustrate the differences in regimental marking that can exist within the same unit when applied after issue from Ordnance stores.**

ment on the barrel – as long as the unit had one and was willing to pay extra. It would seem that this method of marking arms issued from the Tower fell out of general use by the beginning of the 19th century. Few fully regimentally marked India Pattern muskets have been observed, and their markings do not appear to have been applied by the Ordnance Department.

Given that so many muskets going into the hands of Crown forces during the colonial period were not marked at the Tower before issue, it should be no surprise that "in the field" or post-issue markings will be found on many of these arms. Not only do these markings vary in quality and style (from good imitations of Ordnance markings to outright crude and

Advice to Young Soldiers...

"In order to get the character of a smart fellow at exercise, loosen the pins on the stock of your firelock, to make the motions tell. If the piece gets damaged by it, it is no great matter; your captain, you know, pays the piper; and it is right that he should pay to hear such martial music."

"At a field-day, stop up the touch-hole of your piece with cobbler's wax, or some other substance. This will prevent your firing, and save you the trouble of cleaning your arms: besides, unless the quarter-master-serjeant and his pioneers are uncommonly careful, you may secret some cartridges to sell to the boys of the town..."

"In the firings always be sure to fill your pan as full of powder as possible; it will cause much fun in the ranks, by burning your right-hand man: and on the right wing it will also burn the officers; who, perhaps, to save their pretty faces, may order the right-hand file of each platoon not to fire, and thus save them the trouble of dismounting their firelocks, and washing the barrel, after the exercise is over."

"Teach the young recruits the proper use of their arms, when off duty - as, to make a [platform] to hang their wet cloaths upon with the firelocks - with the bayonet to carry their ammunition loaves, toast cheese and pork, and stir the fire: it might otherwise contract rust for want of use."

"In coming down as front rank, be sure to do it briskly, and let the toe of the butt first touch the ground. By this you may possibly break the stock; which will save you the trouble of further exercise that day: and your captain will be obliged to make good the damage. The same effect may be produced by coming from the shoulder to the order, at two motions, especially on the pavement in a garrison town."

Francis Grose, from *Advice to the Officers of the British Army: With the Addition of some Hints to the Drummer and Private Soldier.* First published in 1782, this satirical work gives much insight into the realities of soldiering in the British Army of the Revolutionary War.

(above, left to right) Easily engraved, the brass mounts of muskets were prime recipients for markings. The India Pattern buttplate tang to the left is for number 12 in the Light Infantry Company of the King's (i.e. 4th) Regiment and the two wristplates were marked for issue to E company of the Duke of Rutland's Regiment and G company of the 16th Regiment.

(right) Lots of the fractional marks on British weapons, like the one shown here, have just two parts. Unfortunately, without the crucial third piece of information, these marks tell us nothing about which regiment was issued this piece of equipment.

laughable), they can appear anywhere – even on the triggerguard bow! For these muskets, there is only one rule: all three of the magical numbers (regimental, company and individual weapon number) must be present in order to make a regimental determination. Two-number "fractions" are quite common, but without that critical third component, they are more-or-less useless for identification purposes.

There is one last aspect of regimental markings on Brown Bess muskets that should be made clear. There are a number of Pattern 1742 and earlier muskets that have been observed with extremely high "company" numbers. These markings usually include roman numerals or letters on the top line of the wristplate "fraction." The "company" numbers in these markings are so very high that they could not possibly represent the companies of a British Army regiment. Markings of this type are believed to be American unit markings from the French and Indian War period.

Regimentally marked Brown Bess muskets are particularly intriguing because we can imagine them being carried by specific regiments in battles where those units participated. In some ways, this makes them the most historical Brown Besses. But there is another group of historically inter-

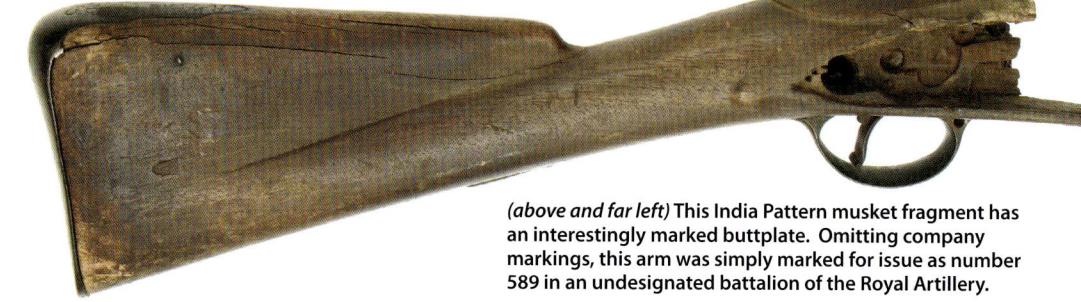

(above and far left) This India Pattern musket fragment has an interestingly marked buttplate. Omitting company markings, this arm was simply marked for issue as number 589 in an undesignated battalion of the Royal Artillery.

esting muskets where our imaginations must play a bigger role. During the Revolutionary War, a large number of British muskets fell into Rebel hands. Many of these weapons were captured at massive British defeats like Saratoga and Yorktown, where whole armies surrendered all at once. Sometimes, captured British soldiers would smash the stocks of their muskets in order to render them useless. Whether damaged or not, these captured Brown Besses provided a flood of locks, barrels and other parts that would be used in American gun construction for years to come. During the war, captured musket parts were a critical asset and were quickly restocked into serviceable arms. Or, like fresh bodily organs illicitly cut from mugging victims in Central Park, these salvaged musket parts could be used to revive broken and unserviceable American arms. After the war, captured Brown Bess parts remained quite common and were readily available to be built into other arms.

The result of all this is that we come across a wide variety of antique muskets today that were built in America from assorted British musket parts. Often, the crown on the lockplate, the regimental designation on the barrel and other prominent indicators of British ownership have been filed off in order to obscure their origins. A musket built in America during the Revolution from captured British parts would be a pretty exciting find – but how do we tell muskets assembled during the war from those made during postwar years?

The short answer is that you probably can't. The inconvenient factor that no one wants to acknowledge is the U.S. Militia Act of 1792. This law required *"that each and every free able-bodied white male citizen...who is or shall be of age of eighteen years, and under the age of forty-five years...shall...be enrolled in the militia* [and] *provide himself with a good musket or firelock, a sufficient bayonet and belt, two spare flints, and a knapsack, a pouch, with a box therein, to contain not less than twenty four cartridges, suited to the bore of his musket or firelock, each cartridge to contain a proper quantity of powder and ball; or with a good rifle, knapsack, shot-pouch, and powder-horn, twenty balls suited to the bore of his rifle, and a quarter of a pound of powder; and shall appear so armed, accoutred and provided, when called out to exercise or into service..."*

So, all of a sudden, just about every service-aged man in the nation was scrambling to acquire a musket. And most of them wanted to do it as cheaply as possible! They didn't need a great musket...just something good enough to pass muster. A lot of these militia muskets were fixed up from old guns that were hanging around; other times, they were quickly assembled from any combination of loose parts that were available. Many of these parts were bits and pieces of captured Brown Besses left over from the war. So how do you tell a crude gun stocked in America during the Revolution from a crude gun stocked in America in order to comply with the Militia Act of 1792?

Well, like beauty, sometimes a musket's history is in the eye of the beholder. Or to borrow the words of an old-time collector we once knew, "If it's my musket, then it definitely dates from the Revolution. But if it's yours and I want to buy it, then I'm sure it's from the 1790s – and it had better be priced accordingly!

(two views) A club-butt New England musket made using Brown Bess parts captured during the Revolutionary War. Longarms of this type were quite common on the Yankee scene until the 1790s. This form of butt carving, in particular, harkens back to the days of the Pilgrims. Some of the brass parts, while being entirely consistent with actual Ordnance parts, appear to be locally made imitations. This makes sense, because while locks and barrels would have been valued and saved, minor brass elements like ramrod pipes might not have been salvaged so carefully. Note the elongated triggerguard finial and the curious profile of the sideplate's tail, which is more angular and less elegantly formed than on genuine British parts.

(far right) Irregularly equipped and haphazardly trained, the militia was often the target of pointed satire.

MARKINGS COMMONLY FOUND ON BROWN BESS MUSKETS

Lock Exterior

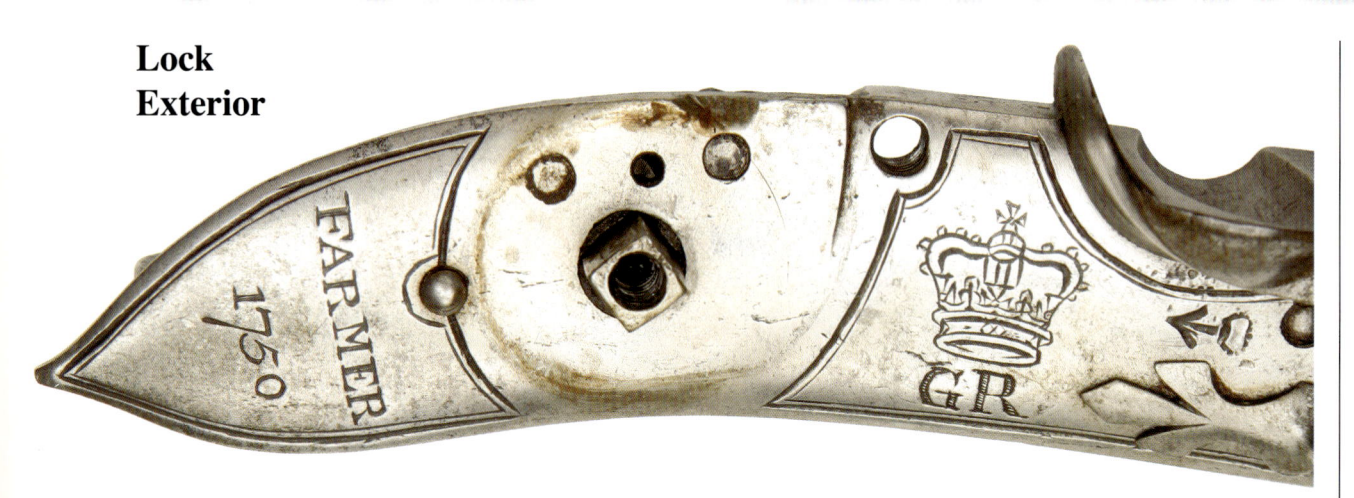

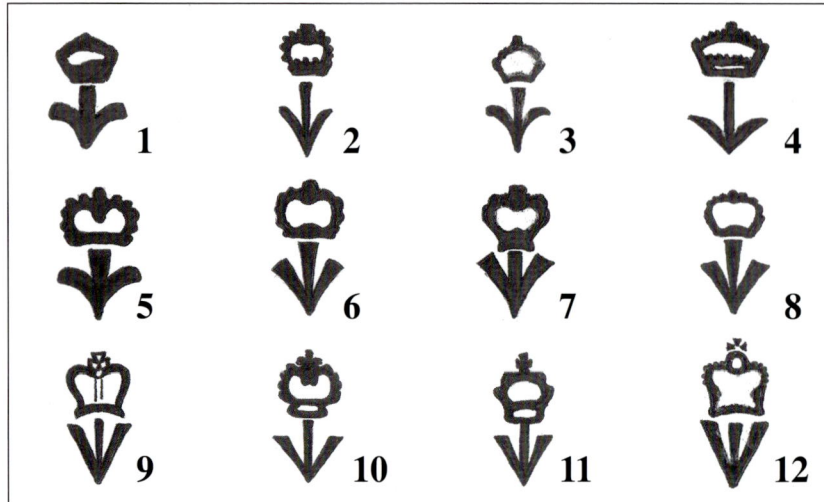

Lock Interior

Inside the lockplate, we usually see two (sometimes more) stamped marks. First, there will be a crown over a number, which is the individual mark of the Ordnance Inspector who accepted the lock. Second, there will be the initials of the person who supplied the lock to the Ordnance. If there is a contractor's name engraved into the tail of the lockplate exterior, then these initials often match those of that contractor. However, if a subcontractor was used, the initials will sometimes not match. Lock interiors will also often be marked with roman numerals. These are assembly numbers that either tie the parts of the lock together or tie the entire lock to the musket itself.

Brass Furniture

Brass parts are often marked. At right, we see the underside of a sideplate with "H" for supplier Thomas Hollier and a broad arrow government ownership mark. At bottom we see a crown inside a triggerguard.

British Ordnance lock exteriors have three essential markings on them:

Engraved at the far left there is the lock contractor's name over the date when it was supplied. This is not necessarily the date when the musket was made, because locks were often stockpiled years before being used. On British muskets, sometimes the word "TOWER" is used in place of the lock supplier's name. After 1764, "TOWER" appears exclusively and the contractors' names and the dates disappear. Irish muskets are usually just marked "DUBLIN CASTLE".

Engraved in the center, there is a crown over a "GR" — the cypher indicating King George. There was always a King George on the throne during the Brown Bess era.

Under the pan, there will be a stamped crowned broad arrow government ownership mark, unless it is a Liège-made lock.

See table *(left)* for dating hints.

Partial lineage of the British crowned broad arrow: **1)** 1706–1711, **2)** 1719, **3)** 1720, **4)** 1727–1729, **5)** 1730–1742, **6)** 1740–1742, **7)** 1743–1745, **8)** 1747–1750, **9)** 1756–1758, **10)** 1756–1760, **11)** 1760–1764, **12)** 1761–1762. Note that Irish Ordnance locks will have their own distinct markings.

14

MARKINGS CONTINUED...

Stocks

(below) Stocks will often be stamped with the musket's assembly number. These roman numerals helped workmen keep the parts of a particular musket together during the manufacturing process. Three common places to find these markings are in the lock mortise, and inside the ramrod and barrel channels.

(below) During most of the period covered in this book, when a musket was assembled, finished and had gone through all of the inspection and acceptance procedures, it was taken into government store and a storekeeper's stamp would be struck into the right side of the butt. This mark took the form of a crowned, addorsed "GR", where the right-side letters mirror the left. At far left, we see the typical mark. Second to the left, we see a variation with a broad arrow added to the bottom. Sometime during the 1780s, dates started appearing under storekeeper's marks, as is shown second from the right. At far right, we see an Ordnance inspector's mark (crown over number) in the ramrod channel — a common mark.

Barrels

Most prominent on Ordnance musket barrels are the King's Proof and View marks (see below). The proof mark is a stamped crown over "GR" over a broad arrow. The view mark is a stamped crown over crossed scepters. Also often present are the barrel supplier's initials and a crowned numeral inspection mark or a simple number. On the tang, we generally see another view mark and (at the very bottom) a crown.

A VERY BRIEF LOOK AT BAYONETS

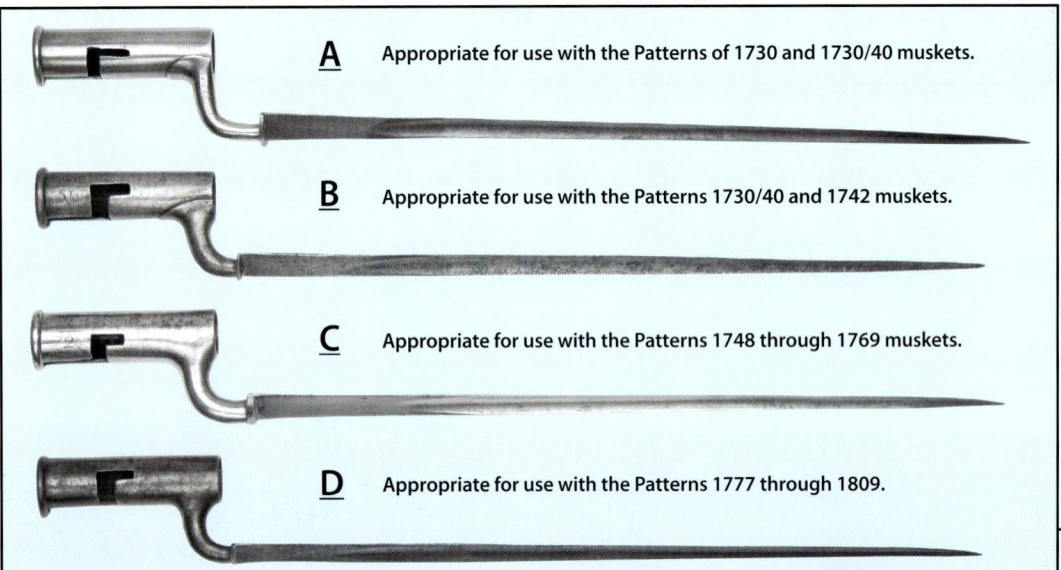

A — Appropriate for use with the Patterns of 1730 and 1730/40 muskets.

B — Appropriate for use with the Patterns 1730/40 and 1742 muskets.

C — Appropriate for use with the Patterns 1748 through 1769 muskets.

D — Appropriate for use with the Patterns 1777 through 1809.

A - Early Land Pattern musket bayonet distinguished by a "shield" shaped decoration at the base of its extremely long shank, c.1720–1735.

B - "Pointed shield" bayonet, c.1735–1745.

C - Land Pattern bayonet without the "shield" and with a wide blade, c.1755–1768.

D - Land Pattern & India Pattern bayonet, c.1768–1810.

These are the four basic patterns of Brown Bess bayonets... but for a more thorough treatment see *The Socket Bayonet in the British Army, 1687–1783* by Erik Goldstein.

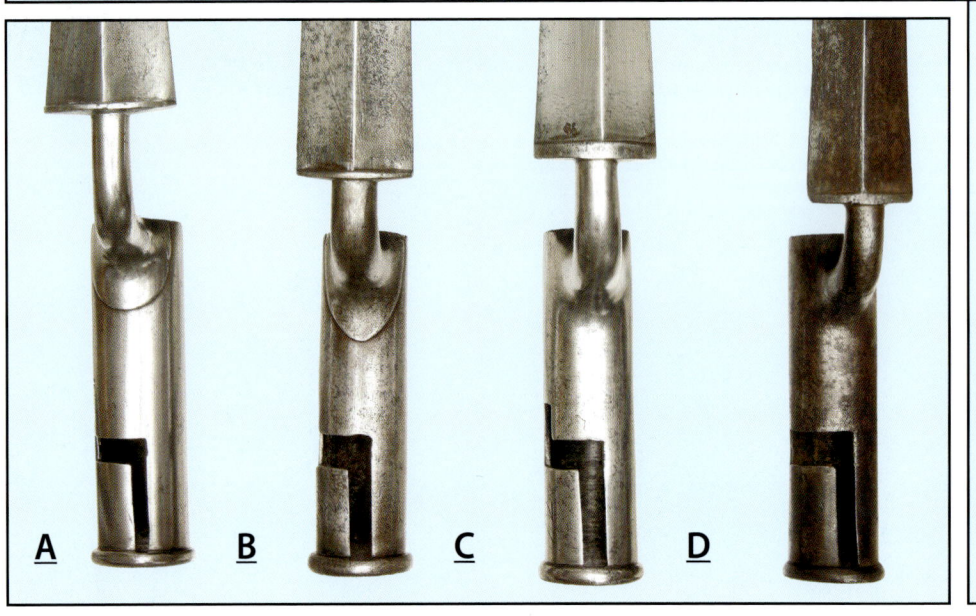

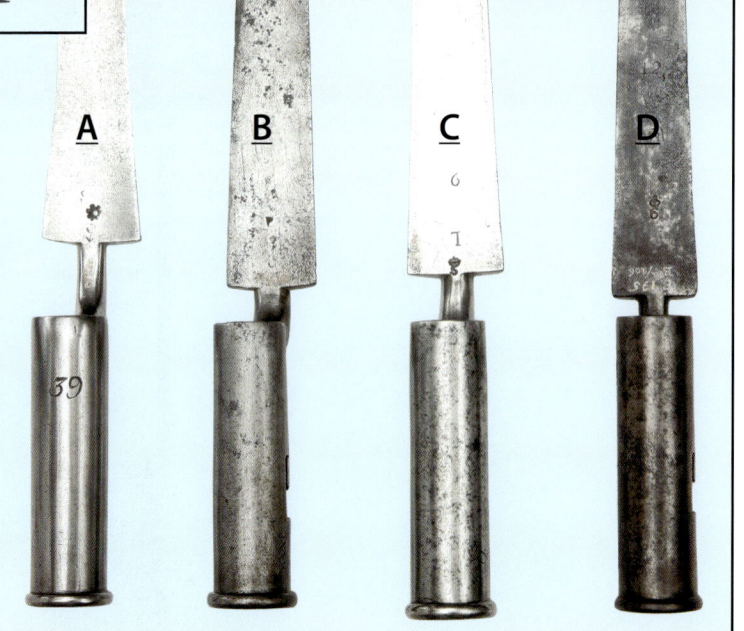

16

A QUICK IDENTIFICATION GUIDE

Experienced Brown Bess enthusiasts tend to identify muskets by their overall look. However, for beginners, a step-by-step system is much more useful. In this guide, we will start with locks. Just like Brown Bess muskets, the locks used in them are identified by Pattern dates. These dates refer to when the lock, itself, was adopted. Sometimes these lock Pattern dates coincide with the Pattern dates assigned to the complete muskets into which they were assembled — and sometimes they don't. There are six basic locks used on Brown Besses. Find out which of these six locks your musket has and then work from there using the hints below the photos.

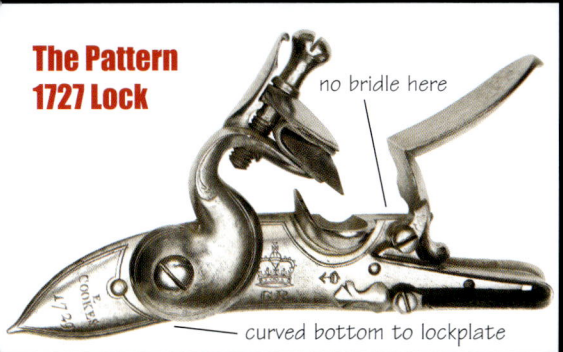

The Pattern 1727 Lock
- no bridle here
- curved bottom to lockplate

This is a fairly simple one. If your lock is a Pattern 1727, then your musket is either a Pattern 1730 or a "Pattern" 1730/40 Long Land Pattern Musket. See the chapters on these muskets (pp. 18 and 30) for details on how to tell them apart. Study all of the photos and text before making any firm decisions. Lots of locks got separated from their original muskets over the years, so pay a great deal of attention to the details of any musket that you are examining. Identifying your lock should point you in the right direction, but it's just a start.

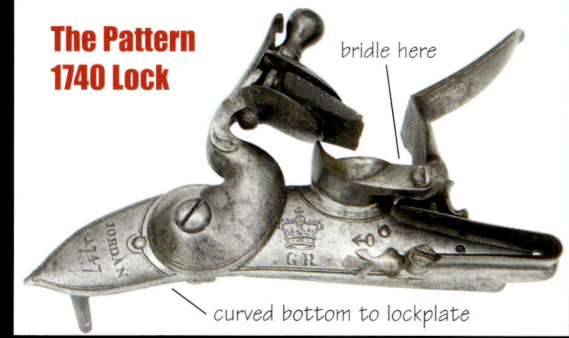

The Pattern 1740 Lock
- bridle here
- curved bottom to lockplate

If your lock is a Pattern 1740, go to the top of page 32 and look at the triggerguard comparison. If your triggerguard looks like the one on the top, then you probably have a Pattern 1730/40 Long Land Pattern Musket. If it looks like the one on the bottom, then there are two choices. Muskets originally equipped with wooden rammers are probably the Pattern 1742 Long Land Pattern; muskets originally equipped with steel rammers are probably Pattern 1748 Long Land Pattern. But it's tricky...so read those chapters.

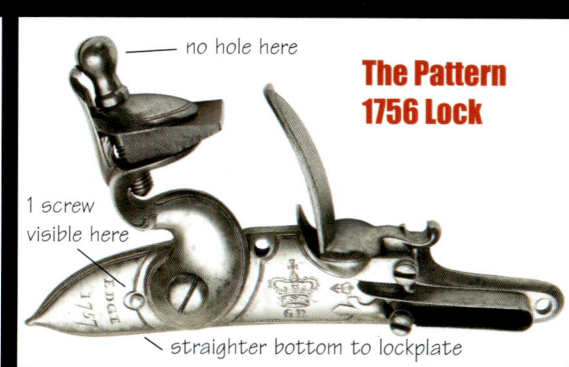

The Pattern 1756 Lock
- no hole here
- 1 screw visible here
- straighter bottom to lockplate

If your lock is an undated Pattern 1756 and the barrel is about 42-inches long, you probably have a Pattern 1769 Short Land Pattern Musket (either British or Irish). If the barrel is about 46-inches long and it has a rear entry pipe for the ramrod (the brass piece that covers the hole where the tip of the ramrod enters the musket's stock), then you probably have a Pattern 1756 Long Land Pattern (British or Irish). If the barrel is about 42 inches long, but that rear entry pipe is missing, you might have a Militia or Marine Musket.

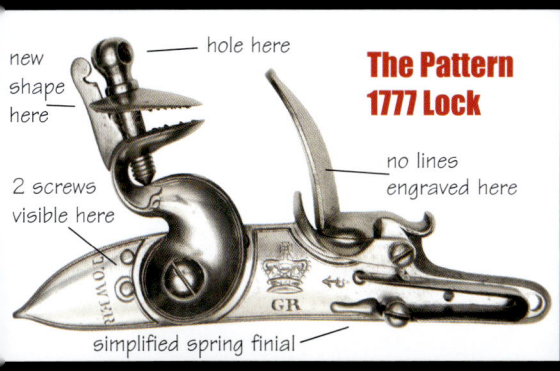

The Pattern 1777 Lock
- new shape here
- hole here
- 2 screws visible here
- no lines engraved here
- simplified spring finial

If your lock is a Pattern 1777 and the musket's sideplate is flat (like the one shown on p. 99), then you probably have a Pattern 1777 Short Land Pattern Musket. If your lock is a Pattern 1777 and the sideplate is not flat at all and looks like the one illustrated on p. 135, then you might have a Pattern 1779-S. If your lock has mixed features, some matching the Pattern 1777 and others matching the Pattern 1756, then you might have Liège Short Land Pattern Musket.

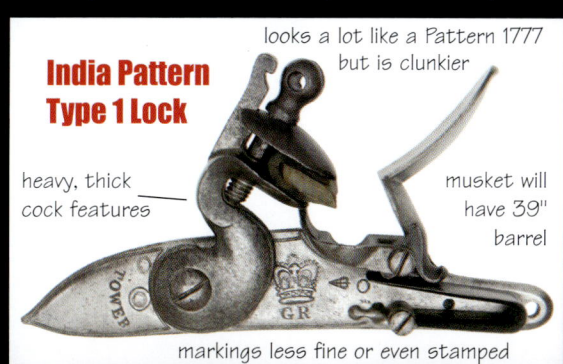

India Pattern Type 1 Lock
- looks a lot like a Pattern 1777 but is clunkier
- heavy, thick cock features
- musket will have 39" barrel
- markings less fine or even stamped

If your lock is an Ordnance-marked India Pattern Type 1 lock, then you almost certainly have a Pattern 1793 Musket, also known as the India Pattern Type 1. While the features of this lock are similar to the Pattern 1777 lock that we just described (see left), the manufacturing quality — or lack thereof — should quickly identify it as an India Pattern weapon. If you are still stumped, be aware that India Pattern muskets have a shorter barrel, measuring just 39 inches.

India Pattern Type 2 Lock
- no notch here
- oval cut out here
- very thick here
- musket will have 39" barrel
- markings usually very flat with almost no depth

If your lock is an Ordnance-marked India Pattern Type 2 lock, you almost certainly have a Pattern 1809 Musket (a.k.a. the India Pattern Type 2). This lock is easy to identify, having a pierced cock with the top-jaw screw going directly down into the hole. Be aware, however, that when musket parts broke, their owners often shoved any old thing in there as a replacement — locks included. Study the detailed photos in this book's chapters carefully to avoid confusion.

Note: If the guide above does not work for a musket that you are looking at, then it is probably either a weapon made from non-matching parts, a musket that was altered in some way, a non-Ordnance weapon, or an intriguing, and perhaps valuable, rarity. Remember, 18th century products are difficult to place in neat categories. There are always exceptions to the rules and totally

THE PATTERN 1730 LONG LAND MUSKET

The classic design that spawned a legend. This elegant pattern is very rare today.

Rarity: Very Rare

Average barrel length: about 46 inches
Average overall length: about 62 inches
Barrel caliber: .76 but loading a smaller ball

In the late 1720s, when the Board of Ordnance finally settled on the pattern that was to become the first standard issue arm of the British soldier, they really hit the ball out of the park – aesthetically speaking. Complete with all the "bells and whistles" musket aficionados love, the Pattern 1730 can be thought of as the '57 Chevy of the Brown Bess series.

A well formed and engraved "banana"-shaped lockplate tricked out only with an internal bridle, raised carving snaking around the breech area, and an exotic triggerguard design borrowed from Europe are some of the markers for this firearm. Another of the more attractive and memorable traits is the seemingly baseball-sized swell at the entry pipe. Some of the other, less-glamorous features include doubled barrel pin loops (designed to hold the tenon of the rammer pipes in between), a high flash fence on the back of the unbridled (and very deep) pan, and a top jaw that approximates the breadth of a half dollar coin.

The design of the firing mechanism incorporated into these fine muskets pre-dates the final musket pattern by three years, and is thus known as the Pattern 1727 lock. Arguably the prettiest martial flint lock ever manufactured, its sole functional drawback was its lack of a bridle connecting the outer edge of the pan to the frizzen's pivot screw. Regardless, no less than seven contractors supplied these locks to the Ordnance during the 1727 to 1738 period. In addition to supplying complete locks, forgings were also sold to the Tower, where they were assembled and engraved with the usual markings, but with TOWER and the date of manufacture on the tail of the lockplate. These seven sources, and their relevant dates are:

Firm	Supply Dates	Observed Dates*	Location
Cole, Elias	1728-1730	1728, 1730	Birmingham
Cookes, Edward	1729	1729	" "
Farlow, John	1728-30, 1737	None	" "
Farmer, Joseph	1727-32	1729, 1731	" "
Jordan, Edward	1734-38	1734	" "
Smith, John	1736-38	None	" "
Vaughan, John	1728-38	None	London
Tower (of London)	N/A	1727, 1728	London

There are sure to be locks not recorded in this chart.

While the first complete Pattern 1730s began rolling into stores at the Tower of London shortly after the date of the pattern, in actuality the Ordnance had been preparing for this specialized production method for a number of years. Musket locks dated as early as 1727 show that components were being stockpiled until sufficient quantities were on hand to begin sending them out for assembly. Thus, it is not impossible that a Pattern 1730 musket could have been originally manufactured with a 1728 dated lock and a post-1735 barrel made with single loops – a simplification made to the pattern at that time.

Struck below the pan on each lockplate should be a "crowned broad arrow," denoting British government property. During the production period of the locks observed on Pattern 1730 muskets, two distinct styles of mark have been observed. The 1727 to 1729 type is characterized by a longer, more delicate "broad arrow" and a straight-sided crown with a serrated arch forming its

Pattern 1730 Long Land Musket

Production Period:
c.1730–1740

Total Production:
Unknown, but significantly less than 96,000

Lock Dates Observed:
1727–30, 1731 & 1734

top. In the 1730 to 1734 period, the mark was much more robust, with a pumpkin-shaped crown composed of a serrated top with sides *en suite*. Appropriately, both styles of mark appear on the Pattern 1727 bayonets made for issue with these muskets.

Regardless of when a Pattern 1730 was made, chances are that they didn't leave the Tower of London until 1740–41, when the Ordnance attempted to rearm as much of the British army as possible with the new, standardized muskets and bayonets. While a specific production tally for Pattern 1730 muskets has not been located, a rough indication can be gleaned from the number of musket bayonets supplied to the Ordnance. Between 1727 and 1740, the years in which Pattern 1730 muskets and their components were produced, slightly more than 96,000 bayonets were provided for "land service." Being that the Ordnance was certainly procuring bayonets for the other arms it was issuing, as well as to replace those lost and broken in the field, the number of Pattern 1730 muskets actually produced is sure to be far less than the 96,000 bayonets contracted for.

This lock, being in pristine condition, exemplifies many of the features that make people fall in love with the earliest Brown Besses. The beautiful form of the lockplate is immediately apparent when viewing this photo. Notice how the tail dips down, displaying a delicate form that is missing from later lock designs. Other early features are indicated below.

Edward Cookes, the lock contractor, and lock manufacture date of 1729

lip

Deep pan, no bridle

The bottom of the frizzen is longer and more graceful than on later models

High-quality engraving

banana-shaped lockplate

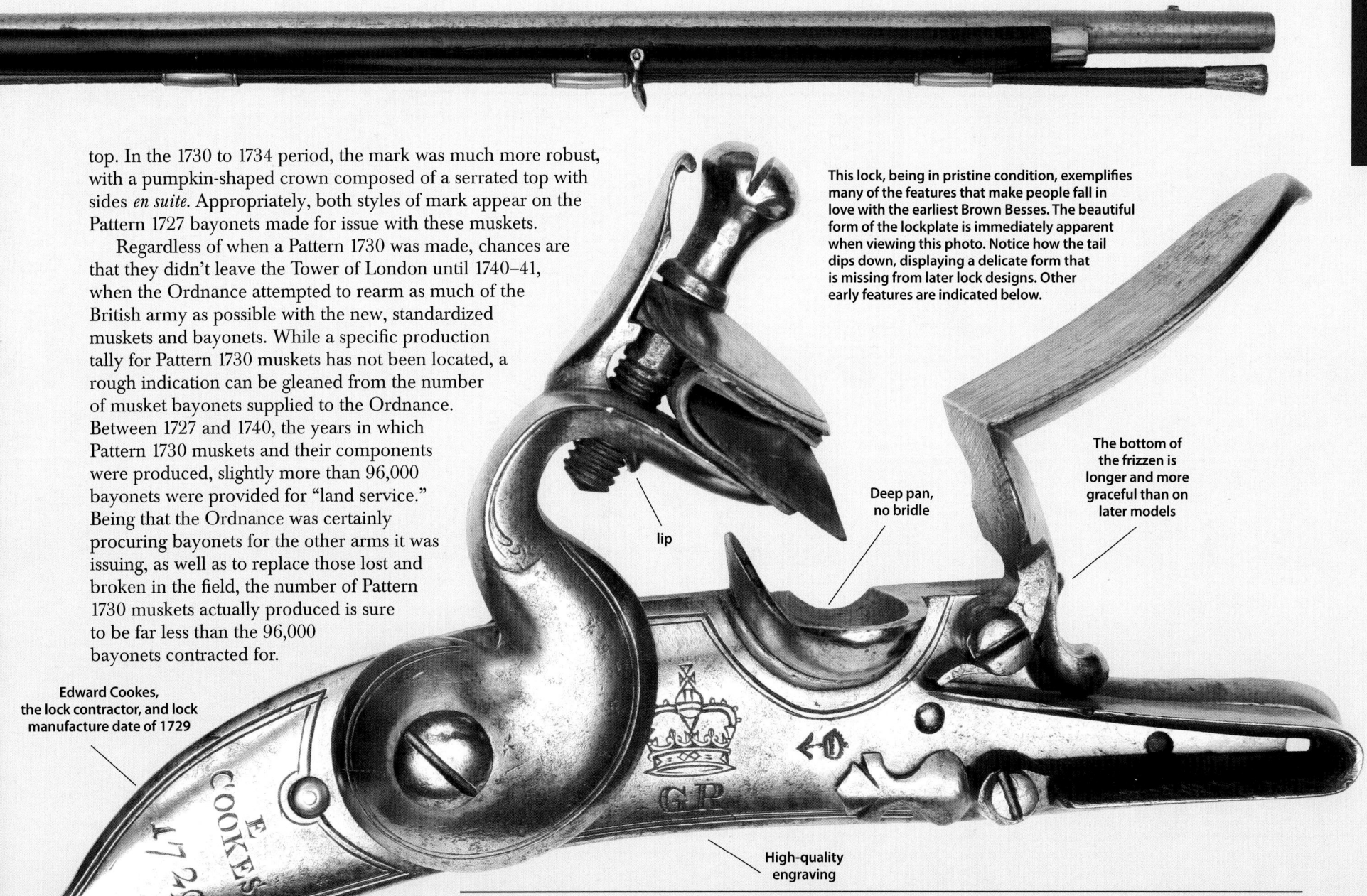

Early-style triggerguard finial.

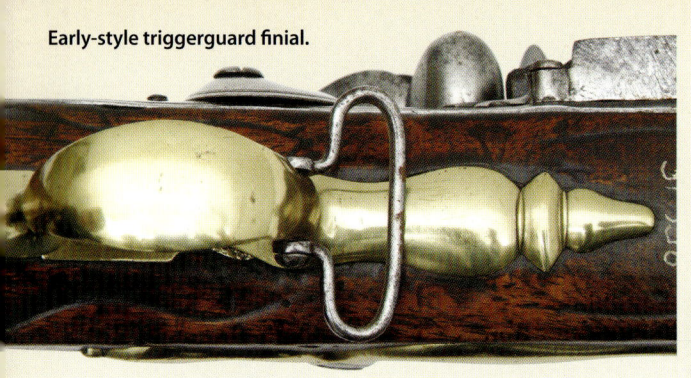

Inherently weak bow on the triggerguard

Regardless of how many tens of thousands of these beauties were produced, they are surprisingly, or perhaps not surprisingly, very rare today. To use what is by no means a scientific method, one could look at the collection of the Colonial Williamsburg Foundation. Probably the most voracious customer for Brown Bess muskets since the Board of Ordnance, the Foundation was especially active in the two decades following World War II, a time when far many more good muskets were available than today. Of the two to three hundred 18th century British military muskets acquired by Colonial Williamsburg, only three true 1730s were included. In other words, only 1 to 1.5% of the total collection! Outside of those few at Colonial Williamsburg and in private collections, examples can be found at the Valley Forge National Historic Park (in the George Neumann collection), the National Army Museum in London, the Royal Armouries in Leeds, and the Regimental Museum of the Royal Welch Fusiliers at Caernarfon Castle in Caernarfon, Wales.

Of the eleven examples studied by the authors, including those at Colonial Williamsburg, eight came from the 18th century arms display at Flixton Hall, Suffolk (UK), and are in very fine condition. Their excellent state of preservation is not surprising since they were discriminatingly chosen and were hung on the wall of a country house where they remained from the 1750s until the 1950s. One other Pattern 1730 studied, certainly from an ancient English arms accumulation, is also in fine original condition and may have also been sprung from Flixton. The three others are in what one might casually call "American" or "attic condition," meaning that they are very well worn, altered and/or patinated. Thus, without the Flixton Hall group, we would know quite a bit less about what these weapons looked like when they were originally issued.

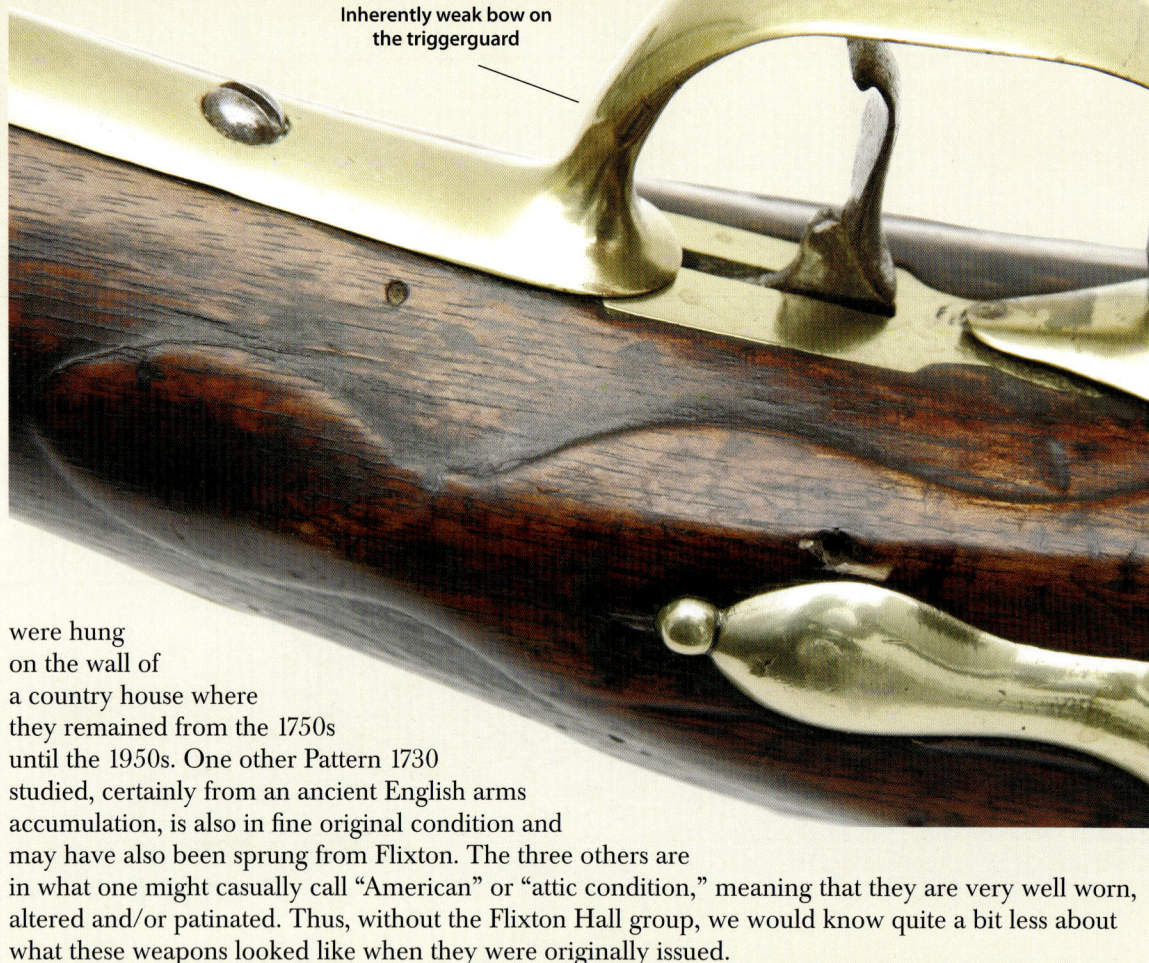

While it is tempting to think that only obsolete or secondary arms were sent from Ordnance stores to the American colonies, such isn't the case here. Before their general debut during the re-arming of the British Army in 1740–41, 700 to 800 Pattern 1730 muskets were shipped across the Atlantic to arm Oglethorpe's (42nd) Regiment in 1737–38. This is surprising, perhaps, but we must remember that Oglethorpe was a personal friend of George II. Sometimes, it's good to know the king!

Perhaps the most dramatic survival of a Pattern 1730 musket sent to colonial America is the relic buttstock recovered from the wreck of the *Machault,* a French frigate sunk by the British in the Restigouche River, Quebec in 1760. Although the iron parts have long since corroded away, the stock is in beautiful condition for wood that has been under water for more than two hundred years. Its five surviving brass mounts include the ornate triggerguard found only on the Pattern 1730 and Pattern 1730/40s, a typical Long Land butt plate, sideplate and triggerplate, and its regimentally marked wristplate. Engraved on the later is the mark "b / 41 / 50th Rt," applied in Boston, where the regiment was raised in 1755. The fact that the bulk of the 50th Regiment capitulated at Oswego in 1756 explains the presence of this musket aboard a French ship in Canada.

Pattern 1730 Long Land Musket

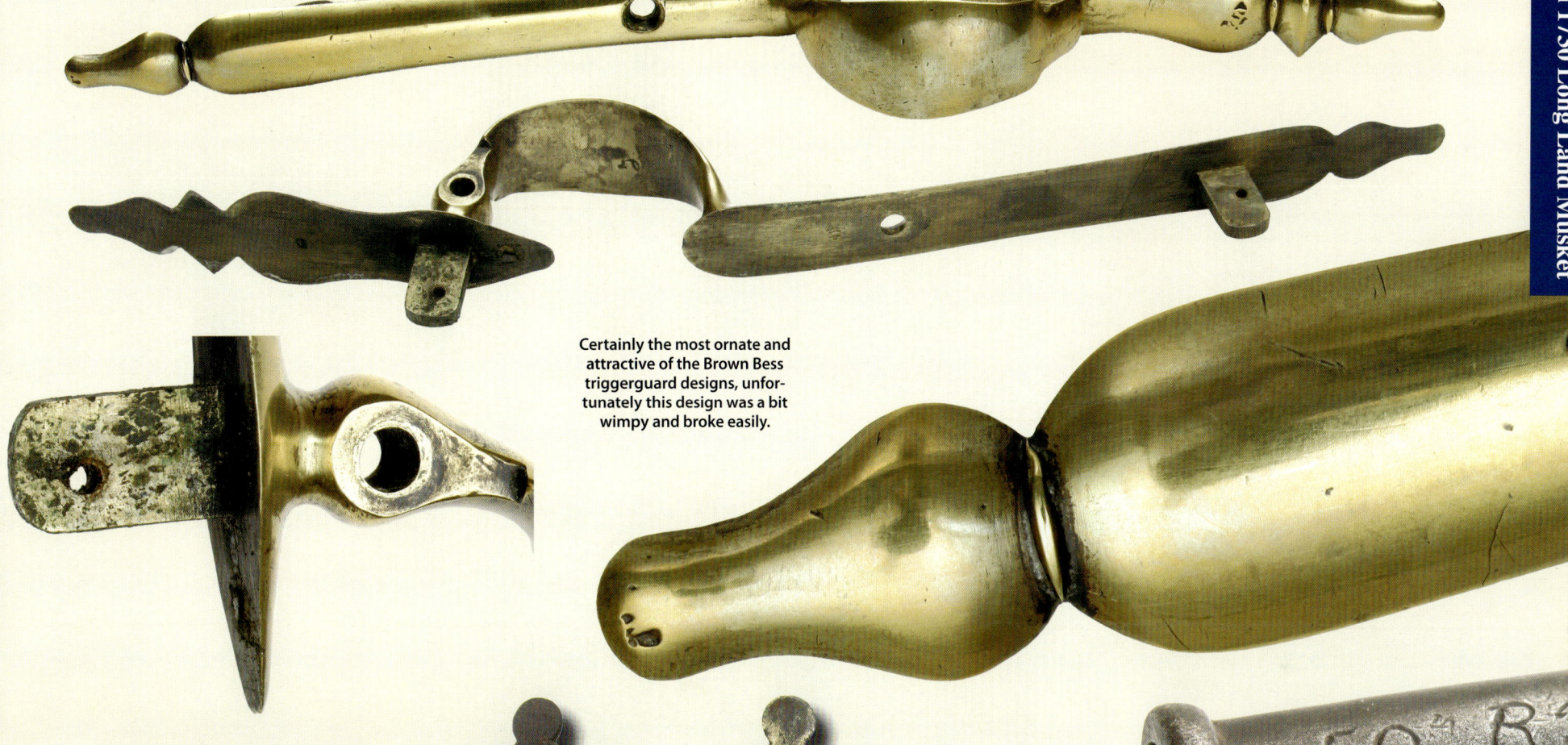

Certainly the most ornate and attractive of the Brown Bess triggerguard designs, unfortunately this design was a bit wimpy and broke easily.

In addition to the *Machault* musket, three other wristplates marked to the 50th have been found, in such places as Fort Ticonderoga and Fort Hendrick in the Mohawk River Valley of New York. Triggerguard finials from Pattern 1730s have also been found at Fort Ticonderoga, and what are likely relics of Oglethorpe's 1737 issue have been found at Fort Frederica, on St. Simons Island, Georgia.

(left and right) Relics of Montcalm's victory at Oswego in 1756, these artifacts of Shirley's 50th Regiment would have been engraved in Boston where the unit was raised.

The two wrist plates are from Pattern 1730 muskets and the bayonet is of the "shield" type described as Type B in the bayonet guide on page 16.

21

The Pattern 1730 had a huge top jaw, approaching the size of a half dollar in girth. We know that the parachute-shaped indentations look like a photo from the WWII Normandy invasion, but they are individually struck cuts meant to hold the flint wrapping (usually lead or leather) firmly in place.

22

Ordnance inspector's mark

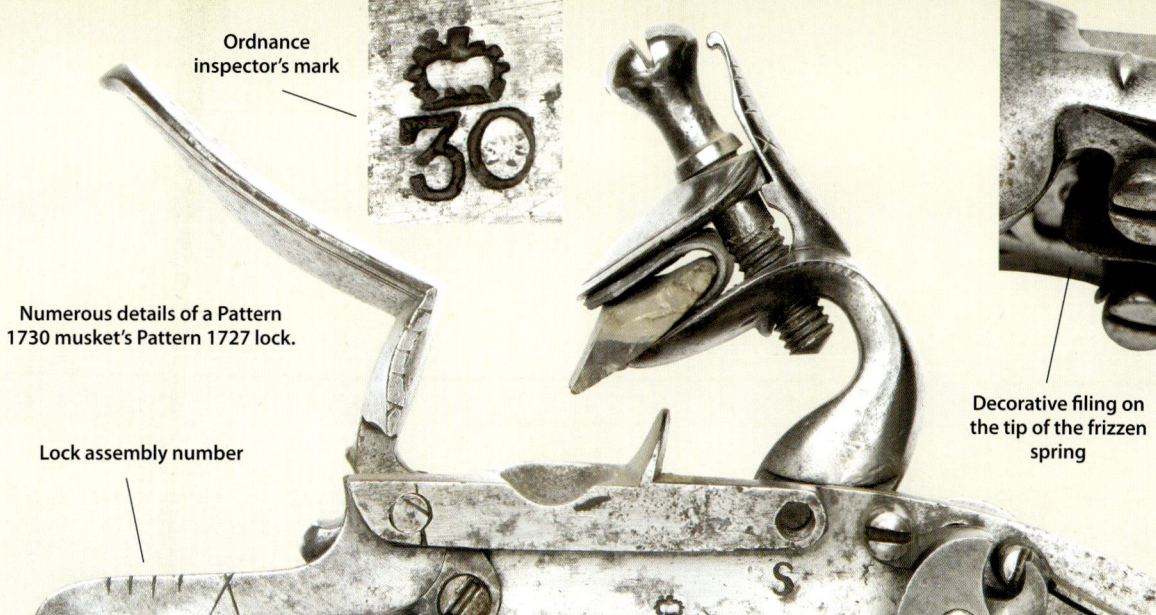

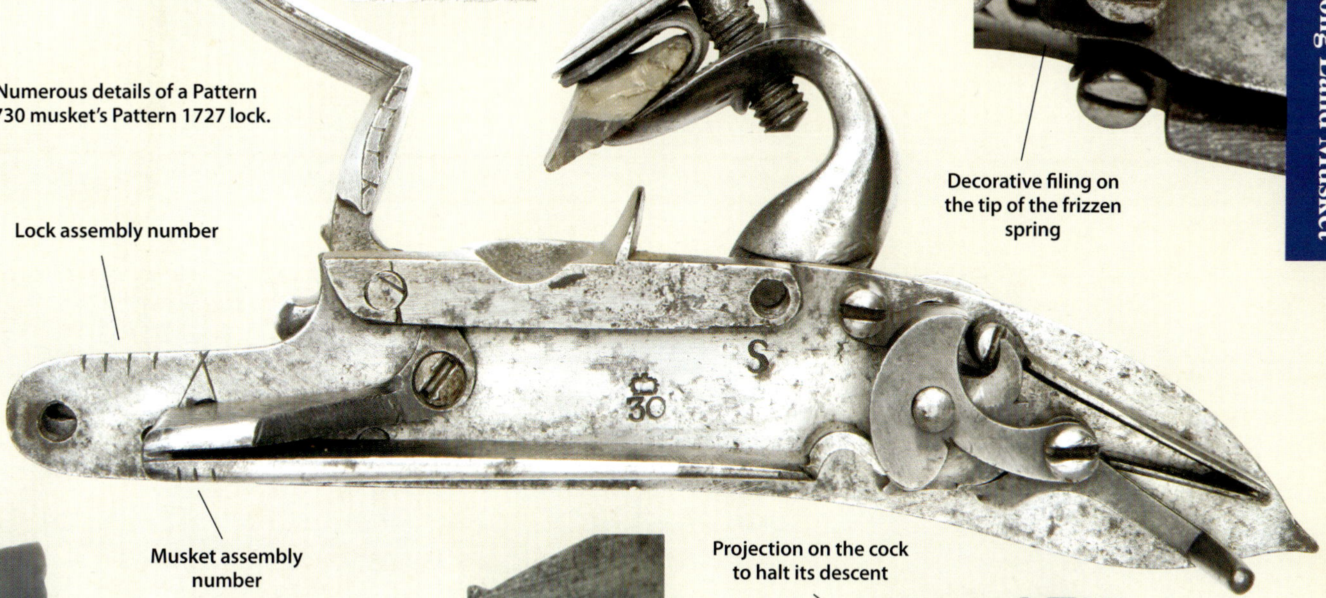

Numerous details of a Pattern 1730 musket's Pattern 1727 lock.

Lock assembly number

Musket assembly number

Deep pan with no bridle

Lockmaker Edward Jordan's mark

Decorative filing on the tip of the frizzen spring

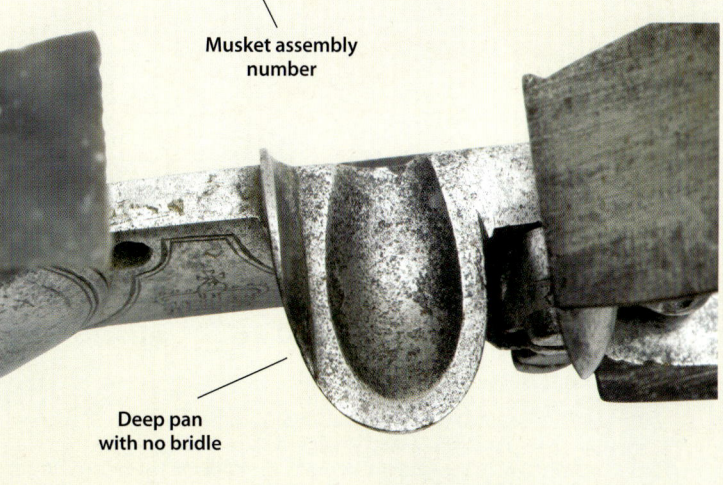

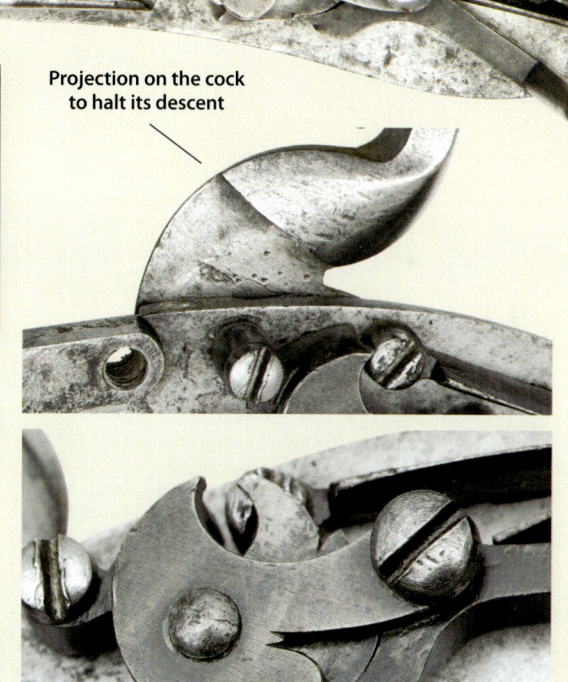

Projection on the cock to halt its descent

Pattern 1730 Long Land Musket

23

Thomas Hollier's maker's mark cast into the underside of the side plate.

Numerous views of the sideplate, showing how it was attached to the musket, sitting in a shallow mortise in the stock. Sideplates from two different examples are shown here. Note the view at the above left, showing a broad arrow mark accompanied by an "H" for Thomas Hollier, the Ordnance's tenant at Armoury Mills until his death in 1754.

If you look closely, you can also see Hollier's mark on the underside of this sideplate, once again accompanied by a broad arrow and, in this case, the roman numeral musket assembly letter "X" (representing 10).

24

The Pattern 1730 musket as a pronounced swell at the entry pipe.

Pattern 1730 Long Land Musket

Inside of the barrel channel, showing the incredibly neat barrel lug mortise. Ordnance contractors certainly did excellent work. Also note the musket assembly number on the left side of this photo.

(left) Ramrod channel marking, showing what appears to be two crescent-shaped marks and a crowned 14 Ordnance inspector's mark.

(right) The "IK" initials in the ramrod channel. These initials probably represent the contractor who set up this musket. These contractors often had large shops, so it would be incorrect to assume that he necessarily did all the work himself.

Seen from the side, the stock shows a distinctive and severe swell in the area of the entry pipe. Not merely attractive, this feature was meant as a grip to help the soldier handle his weapon.

Lock mortise of a Pattern 1730 musket. This mortise was intended to accept the lock as tightly as possible while retaining the most wood possible for reenforcement of the stock.

(below) This sheet-brass nose band was added at the regimental level after it was issued by the Ordnance. It was intended to stop the forestock from splitting.

Triggerguard mortise, very shallowly cut, showing the ornate shape of the finial. This mortise barely penetrates the surface of the stock.

More neat mortise work.

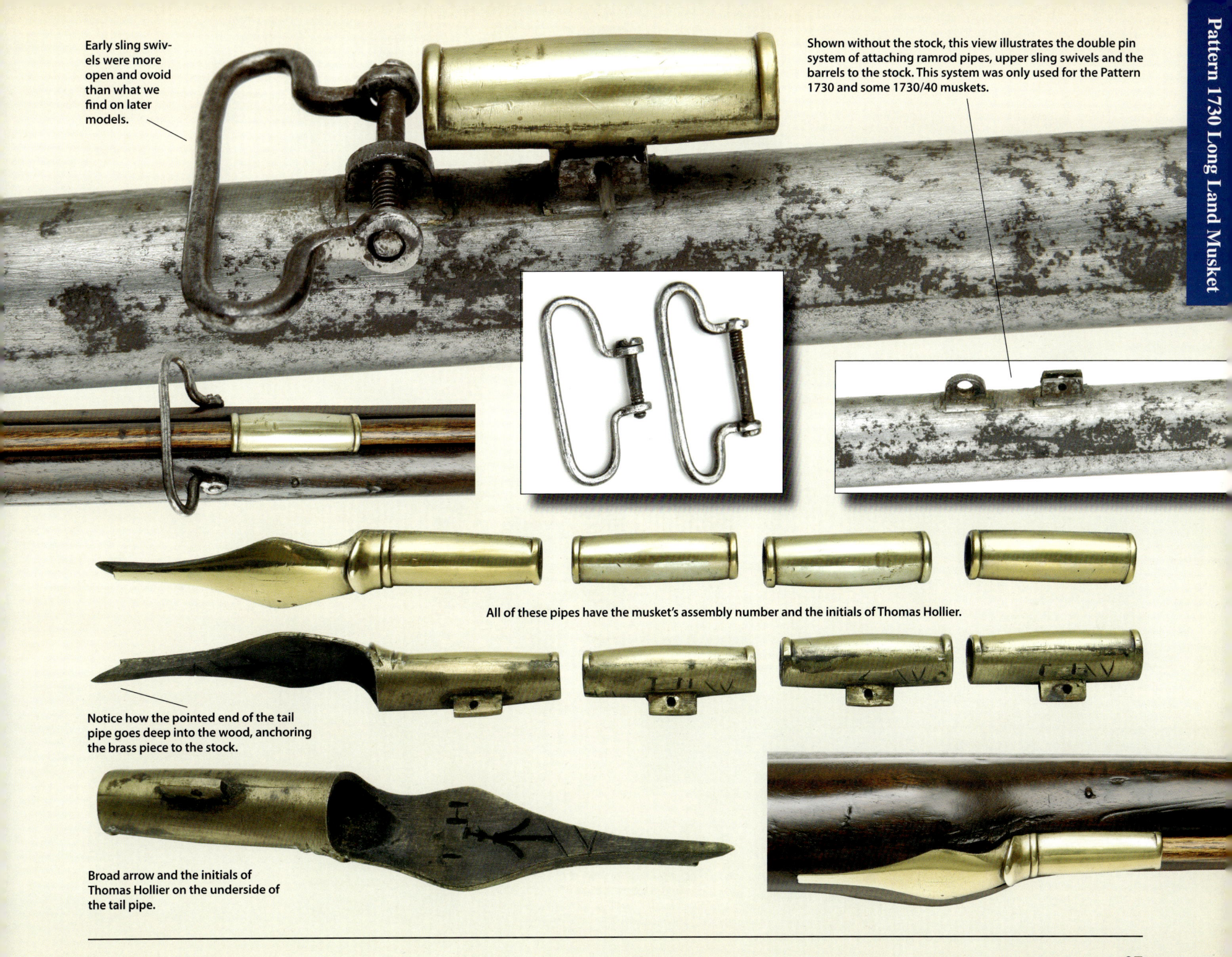

Pattern 1730 Long Land Musket

Early sling swivels were more open and ovoid than what we find on later models.

Shown without the stock, this view illustrates the double pin system of attaching ramrod pipes, upper sling swivels and the barrels to the stock. This system was only used for the Pattern 1730 and some 1730/40 muskets.

All of these pipes have the musket's assembly number and the initials of Thomas Hollier.

Notice how the pointed end of the tail pipe goes deep into the wood, anchoring the brass piece to the stock.

Broad arrow and the initials of Thomas Hollier on the underside of the tail pipe.

27

(above) Breech area of the barrel, showing Regimental designation, Tower view and proof marks.

(above) Ramrod tip. Original wooden ramrods on Brown Bess muskets are exceedingly rare.

(above and below) Another of the salient features of the Pattern 1730 is the graceful, well-shaped comb of the buttstock which terminates in a faceted notch.

Ordnance storekeeper's mark

(left) Musket assembly marks on the barrel. Also note that the decorative filing on the end of the breech does not extend into the part of the barrel that is hidden by the stock.

(below) Crowned crossed scepters. These are the view marks from when the barrel was proofed, indicating that the breech assembly was deemed acceptable to the Ordnance.

28

Pattern 1730 Long Land Musket

Wrist plates were held in place with a screw that came through the triggerguard and the stock. Because of this screw, the stock was further weakened at this point.

(left) Wrist plate marking indicating that this was the second musket in Company C. The barrel is marked to the Royal Welsh Fuziliers.

(above) Only appearing on the Patterns 1730 and some 1730/40 muskets, the ornate carving around the lock, barrel tang and sideplate mortises are thought to be particularly attractive.

(right) Another wrist plate from a Royal Welsh Fuziliers-issued Pattern 1730 musket, this example being the 39th musket in Company H.

29

THE PATTERN 1730/40 LONG LAND MUSKET

A Pattern 1730 musket upgraded with some features of the Pattern 1742.

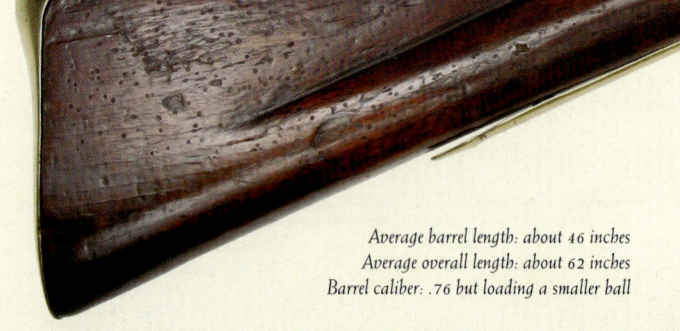

Rarity: Very Rare

Average barrel length: about 46 inches
Average overall length: about 62 inches
Barrel caliber: .76 but loading a smaller ball

In stark contrast to the previous chapter, with its neat and tidy explanation of the Pattern 1730, this chapter is possibly the muddiest in the book. That is because the Pattern 1730/40 musket isn't really a "pattern" at all. It's simply a convenient term used by modern arms students to describe something that isn't quite a "straight up" 1730, due to one improvement or another.

As was described earlier, the Ordnance system of supply began in the 1720s with the stockpiling of components, which was then followed by the stockpiling of assembled muskets. This was done very deliberately, looking forward to a time when the entire British army could be uniformly re-equipped – more or less all at once. Very few of these new muskets were issued until there were enough to go around. As a result of this policy, more than a decade passed between the manufacture of the first Pattern 1727 locks and the general re-equipping of the army with the new "standard" arms.

This delay in issuing finished weapons had the unintended effect of denying Ordnance the opportunity to undertake large-scale service trials of their new Pattern 1730 Long Land Muskets. So if there were any defects in the muskets that needed to be corrected, or improvements that might be suggested by officers in the field, Ordnance would not learn about these issues until it was too late to do much about it.

When large numbers of these new arms went into service early in the 1740s, problems requiring correction manifested themselves pretty quickly. The 1730 Pattern was adjusted to address these concerns. The most important of these changes concerned the performance of the lock; another change addressed a structural weakness, while the third was a cost-saver. If any of these upgrades are seen on what at first appears to be a Pattern 1730, then this musket is referred to as a "Pattern 1730/40" musket by collectors. Distinguishing traits to look for include a double-bridled lock, and/or a less ornate triggerguard, and/or the lack of full Pattern 1730-style stock carving. All of these new features would later become part of the Pattern 1742 musket.

Mechanically most significant, we've all been hearing about the superiority of the double-bridled lock over the single-bridled version for years – and this is certainly true. The performance of a flintlock is directly affected by the ease and rate at which the steel (or frizzen) flies back when struck by the flint, allowing the ensuing shower of sparks to land in the powder-filled pan. A wobbly frizzen bound by friction would flip increasingly more slowly as the lock wore in service, drastically affecting its operation. This problem was remedied by the addition of a "bridle" connecting the steel pivot screw to the edge of the pan, which necessarily raised the cost of the lock. The little metal tab had the desired effect, since an axel supported at both ends always turns truer than one supported at only one end (like a lathe, for instance). While double bridled Land Service musket locks have been observed (albeit surprisingly) with dates as early as 1737, most Pattern 1730/40 muskets will have locks dated between 1740 and 1742.

Back when the Ordnance settled on the final design of the Pattern 1730, it was only natural that they created a brass version of the wrought iron triggerguard commonly encountered on martial firearms, both of British and Continental manufacture. Although heavier than iron, cast brass is far weaker, being a softer metal and created without the strengthening "grain" found in the former material. If one considers that most Pattern 1730s finally went into service around 1740, and a new triggerguard design was officially adopted in 1742, it would seem that it did

30

Pattern 1730/40 Long Land Musket

Production Period:
c.1740–1742

Total Production:
Unknown

Lock Dates Observed:
1740–1742

not take long for the Ordnance to find out that their original triggerguard design was a complete failure – regardless of how attractive it was.

If one is lucky enough to have access to muskets with both types of triggerguards (1730 and 1742), a side-by-side comparison will make the differences between them obvious. The earlier, fancy version is notably slighter, with a narrow bow flimsily connected to the two anchoring straps. In sharp contrast, the "1742" type is much broader and beefier, with an extra, incurving curl behind the trigger. While this feature isn't connected to anything at its terminus, its purpose is not solely decorative. This curl acts as a brace to prevent the bow from being crushed down against the trigger, thus immobilizing it and rendering the musket un-fireable. Just how much heavier was this new "1742"-style triggerguard, you might ask?

Firm	Supply Dates	Observed Dates*	Location
Clarke, William	1741-42	None	London
Farmer, James	1741-42	None	Birmingham
Jordan, Edward	1740-42	None	" "
Jordan & Farmer**	1742	None	" "
Smith, John	1740-42	None	" "
Vaughan, John	1740-41	None	London
Tower (of London)	N/A	1740-42	London

*There are sure to be locks not recorded in this chart.
**Marked with either name, not both.

Pattern 1740 lock.

Pan with bridle

(right) Close-up of bridle, and the new simplified end on the frizzen spring.

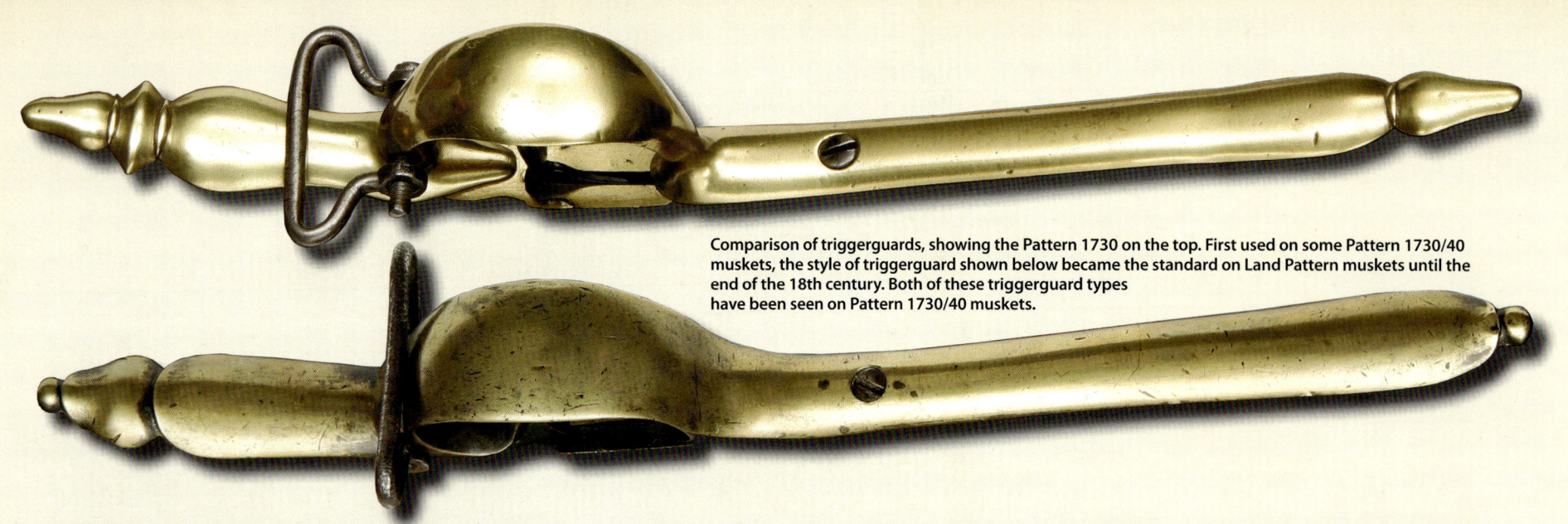

Comparison of triggerguards, showing the Pattern 1730 on the top. First used on some Pattern 1730/40 muskets, the style of triggerguard shown below became the standard on Land Pattern muskets until the end of the 18th century. Both of these triggerguard types have been seen on Pattern 1730/40 muskets.

Thanks to an accurate scale and the necessary firearms at Colonial Williamsburg, it can be said that (on average) the 1742 pattern triggerguard was about 33% heavier than its predecessor, proving that substantial reinforcement was added to the design change.

In addition to the newly placed pan bridle and strengthened triggerguard, Land Service locks produced in the late 1730s and early 1740s also exhibit some minor changes. Not affecting the function of the mechanism at all, these seemingly trivial alterations were apparently implemented to save money, and are not obvious to the casual observer. Top jaws had their diameter reduced by a few millimeters, while their screws seem to have lost some of their tapering grace, and perhaps a bit of their breadth too. Also deleted was that funny little reinforcing lip at the front of the screw hole where it pokes through the bottom of the cock's lower jaw.

As eye-catching as we may think the raised carving around the lock and breech areas of the Pattern 1730 is, it really adds nothing to the musket other than cost. Thus, it was an obvious choice to be deleted from the current pattern in the name of savings – an economy which might have covered the cost of the new, more robust triggerguard and/or the addition of the exterior bridle to the lock.

What seems to be yet another casualty of making these service muskets more affordable in the name of improvement was the double-looped barrel and rammer pipe retention system. While neat in appearance, it was almost certainly a more labor intensive feature, which disappeared during the short production life of the Pattern 1730/40. Examples of both single-looped and doubled-looped arms of this truly transitional type have been studied.

And now for the caveat. Since it has been established that the Pattern 1730/40 must exhibit a combination of components found in previous and future patterns, there is indeed a pitfall. During the period a musket was in use, components – even locks – were often replaced. What may seem to be an honest 1730/40 might just be a Pattern 1730 with a newer lock or a Pattern 1742 with an older one. So, as always, buyer beware!

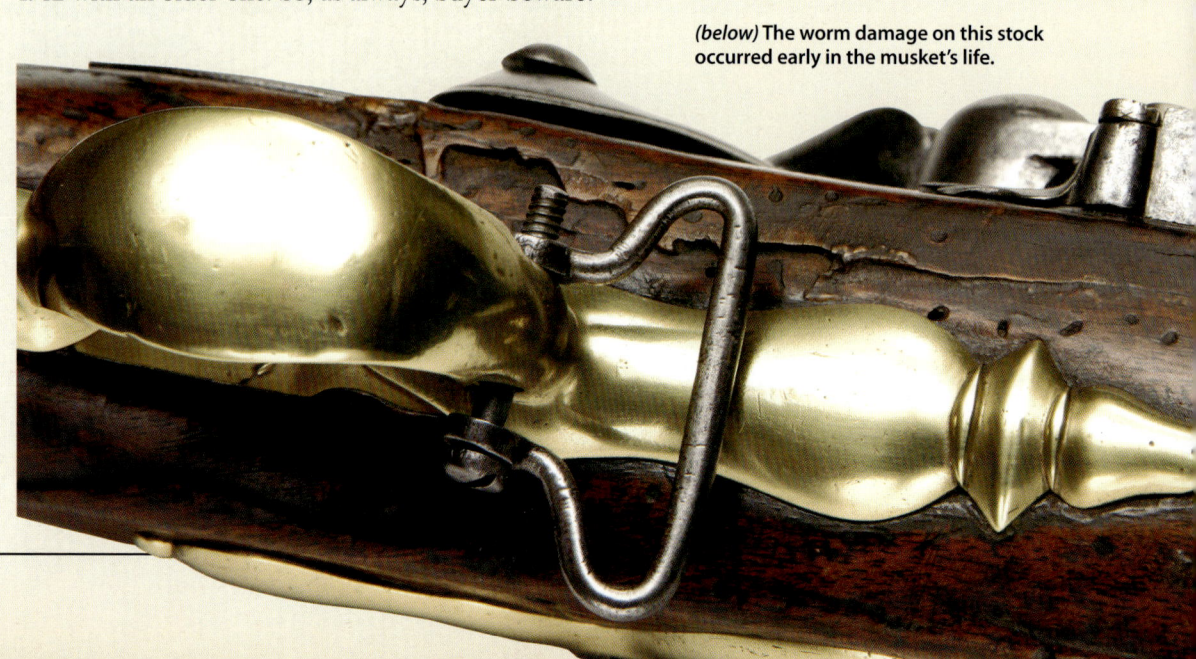

(below) The worm damage on this stock occurred early in the musket's life.

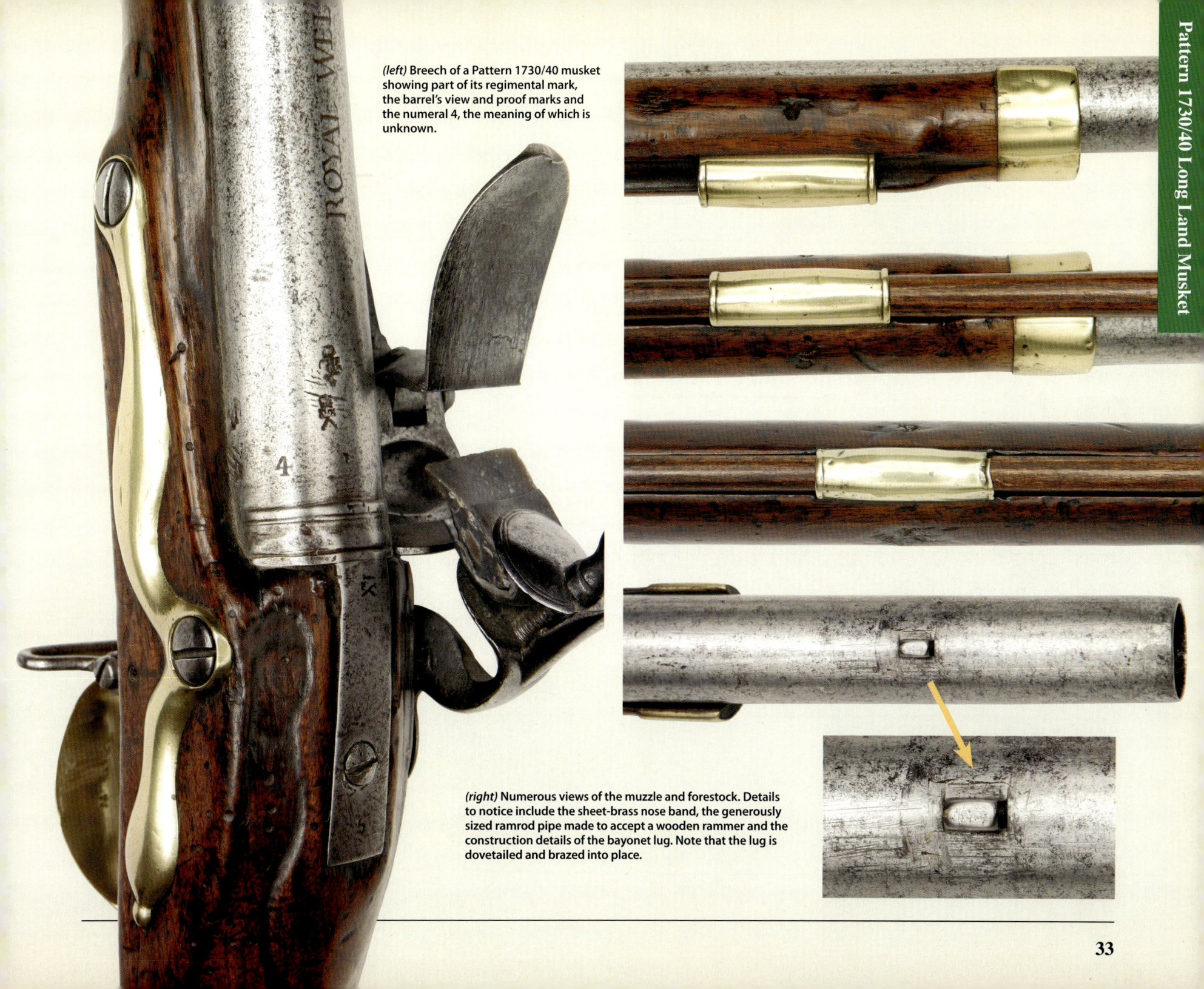

(left) Breech of a Pattern 1730/40 musket showing part of its regimental mark, the barrel's view and proof marks and the numeral 4, the meaning of which is unknown.

(right) Numerous views of the muzzle and forestock. Details to notice include the sheet-brass nose band, the generously sized ramrod pipe made to accept a wooden rammer and the construction details of the bayonet lug. Note that the lug is dovetailed and brazed into place.

Pattern 1730/40 Long Land Musket

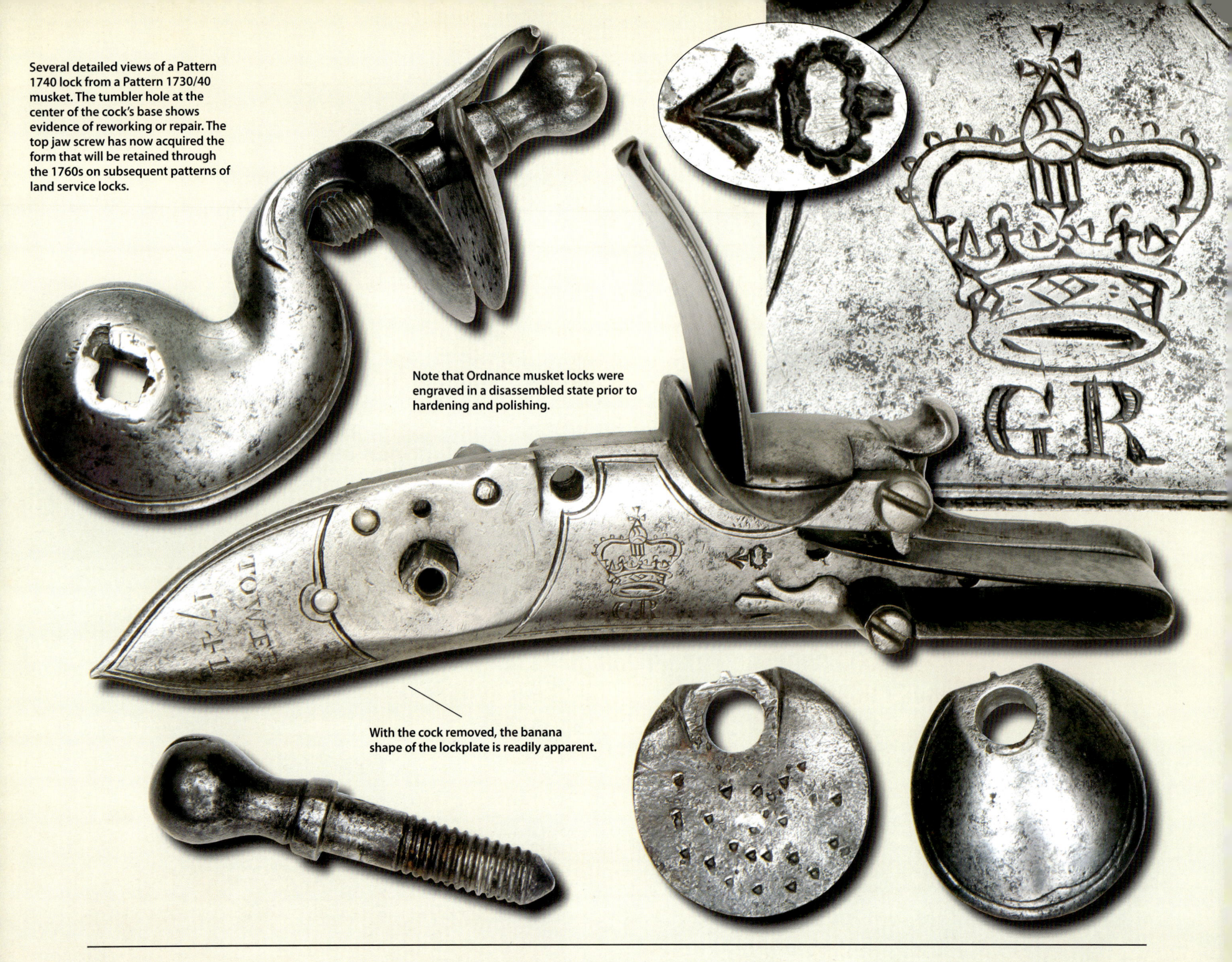

Several detailed views of a Pattern 1740 lock from a Pattern 1730/40 musket. The tumbler hole at the center of the cock's base shows evidence of reworking or repair. The top jaw screw has now acquired the form that will be retained through the 1760s on subsequent patterns of land service locks.

Note that Ordnance musket locks were engraved in a disassembled state prior to hardening and polishing.

With the cock removed, the banana shape of the lockplate is readily apparent.

34

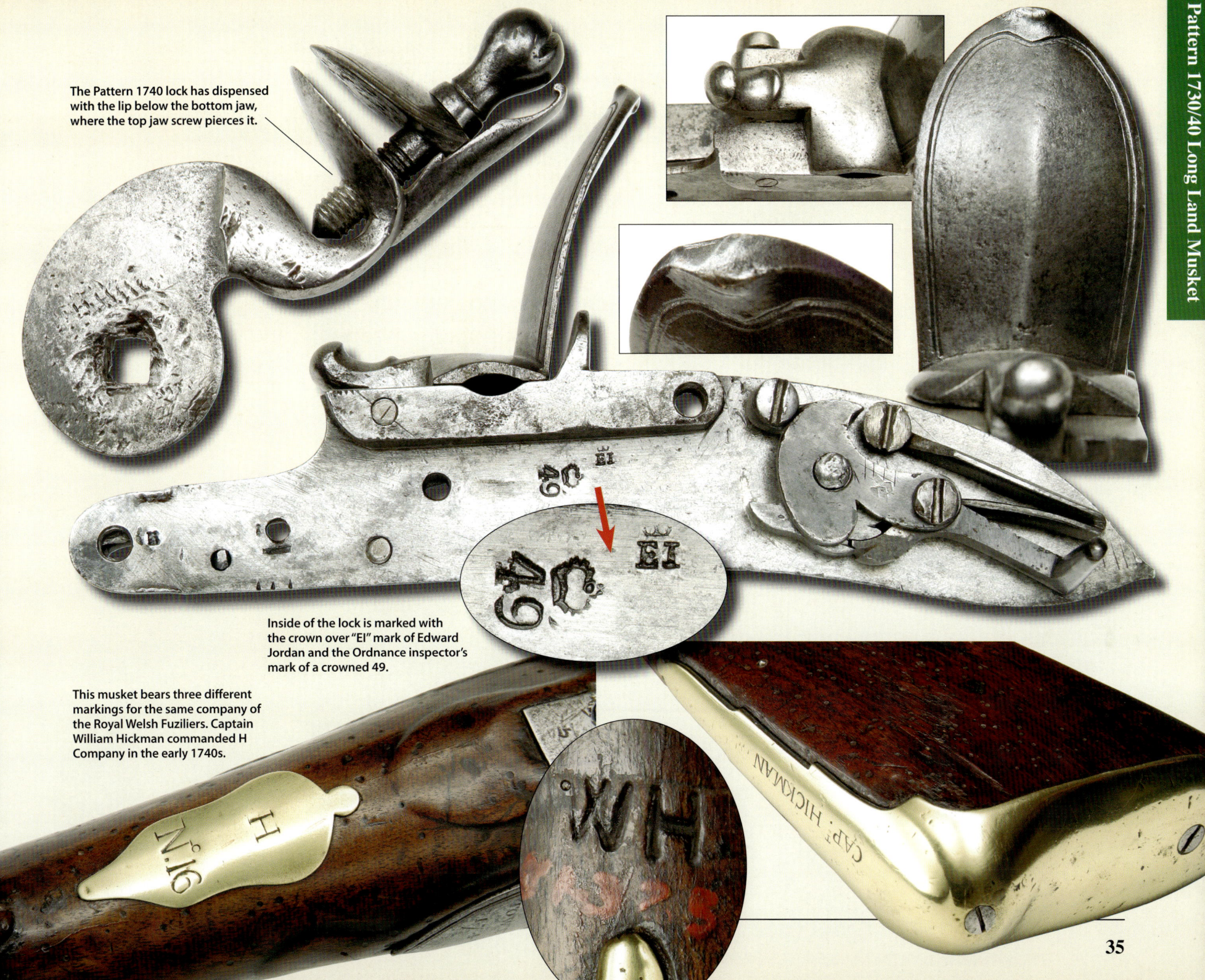

Pattern 1730/40 Long Land Musket

The Pattern 1740 lock has dispensed with the lip below the bottom jaw, where the top jaw screw pierces it.

Inside of the lock is marked with the crown over "EI" mark of Edward Jordan and the Ordnance inspector's mark of a crowned 49.

This musket bears three different markings for the same company of the Royal Welsh Fuziliers. Captain William Hickman commanded H Company in the early 1740s.

35

Details of the stock.

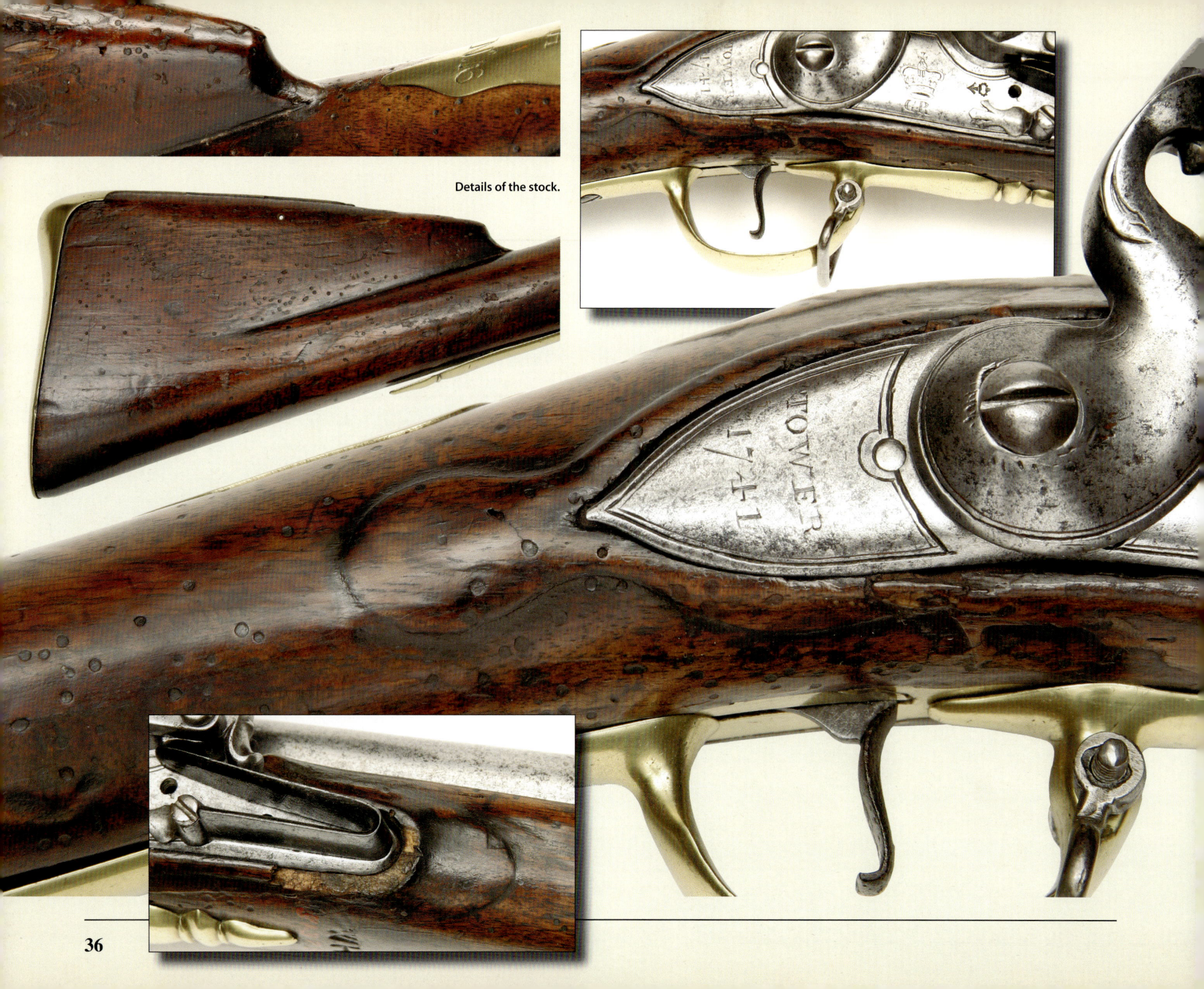

Still deemed useful by the Board of Ordnance, the bulge in the stock at the tailpipe was retained into the 1740s. It is a particularly dramatic and desirable feature on early Brown Bess muskets.

Pattern 1730/40 Long Land Musket

37

THE PATTERN 1742 LONG LAND MUSKET

Simplified stock carving, a double-bridled lock and a heavier triggerguard identify this pattern.

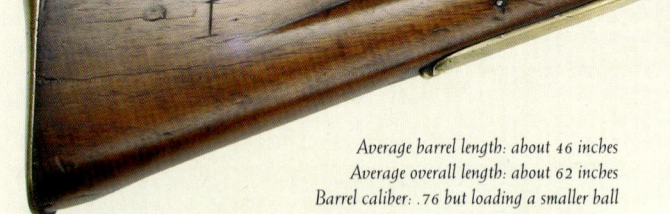

Rarity: Rare

Average barrel length: about 46 inches
Average overall length: about 62 inches
Barrel caliber: .76 but loading a smaller ball

As we move forward in the lineage of the Brown Bess, we are beginning to emerge from the darkness of extreme rarity. While one could probably count all of the surviving Pattern 1730s on both hands and a few toes, quite a few more Pattern 1730-40s survive. However, it is not until we reach the Pattern 1742 that a truly accessible Land Service musket enters the picture, and it's just in time for one of the most important periods in Anglo-American history: the Seven Years', or French and Indian War.

It seems that fighting on the American front consumed or utilized, in one way or another, approximately 30% of the total production of this pattern. Although steel-rammered muskets began arriving at the Tower of London in the late 1740s and went into production high gear in the second half of the 1750s, none of these state-of-the-art patterns were shipped over for this conflict. In period documents, the Pattern 1742 musket can be identified as the most commonly issued Ordnance musket of the war, identified by terms such as "Land Service Musquets of the King's Pattern with Brass Furniture Double-bridle Locks, (and) Wood Rammers." Today, the Pattern 1742 is by no means a common weapon. But there are enough available for study, most of them being well-worn specimens with a few in what one might call "excellent" condition.

Physically, the Pattern 1742 is the "Mack Truck" of the Land Service musket series. Compared side-by-side with a Pattern 1730, one notices that the stock architecture has been bolstered substantially, most notably at the forestock and breech. When viewing the lock areas on these two patterns, the increase in wooden mass between the bottom of the lockplate and the front triggerguard finial is immediately apparent. Similarly, the comb of the buttstock has been beefed up, and is more bulbous in cross section than the Pattern 1730. Clearly, the graceful but slightly svelte stock of the Ordnance's inaugural attempt proved too weak for service, so they allowed Bess to gain a bit of weight during the inter-pattern era of 1740–1742. However, the large swell at the entry pipe remained, albeit a bit less pronounced.

Firm	Supply Dates	Observed Dates*	Location
Clarke, William	1742	None	London
Farmer, James	1742–51	1745–48	Birmingham
Jordan, Edward	1742–50	1744, 1746	" "
Jordan & Farmer**	1742–44	None	" "
Smith, John	1742–48	None	" "
Vaughan, John	1742–47	None	London
Wood, John	1748	1748	London
Tower (of London)	N/A	1742	London

There are sure to be locks not recorded in this chart.
**Marked with either name, not both.*

The new double-bridled Pattern 1740 locks built into these arms are more uniform in appearance than the single-bridled Pattern 1727 locks, and dispense with some of the latter's niceties. Gone is the funny lip on the bottom jaw of the cock, as is the careful shaping at the end of the steel spring and the light-handed, graceful engraving. Since we are still in a period of hand production, there are noticeable differences in locks produced by the various

Pattern 1742 Long Land Musket

Production Period:
c.1742–1750

Total Production:
Unknown, but less than 106,900 (based upon bayonet production)

Lock Dates Observed:
1742–1748

Totals Shipped to America, 1754–1764:
for and with British troops, at least 15,833; for American use, at least 14,798

contractors. For instance, the locks produced by Farmer during the 1746 to 1750 period will often be found with a "chubbier" cock than what is seen on others made during the same period. Similarly, Farmer's locks from this time have been noted with an elongated center tyne to the steel spring finial (see red arrow, below right). Further, detailed study of locks by different contractors would surely illuminate additional idiosyncrasies.

With more than 30,000 Pattern 1742s shipped across the Atlantic during the Seven Years' War, it can be said that this pattern composed the first real influx of standard British martial arms to reach America. Therefore, it isn't surprising that most of the surviving specimens exhibit some form of "Americanization," be it a repair, an update or a marking. All of these alterations should be considered interesting and studied within their contexts, instead of being "restored" in a senseless effort to create a "perfect" archetypal musket.

During this period, British Army structure (as applied to the tracking of firearms) was generally broken down in decreasing order by Regiment, then by company and finally by individual weapon number. While full British regimental markings are seldom encountered on Pattern 1742 muskets, company and weapon number designations (in the form of so-called "rack numbers") are occasionally found engraved on the wristplates.

All Pattern 1742 muskets were originally equipped with Pattern 1740 locks.

Typical for firearms that saw use during the French and Indian Wars, this lock exhibits a great deal of wear.

The threading on the top jaw screw is coarser than normal on this example. It is clearly an Ordnance part, but perhaps one that was rethreaded in order to extend its service life.

39

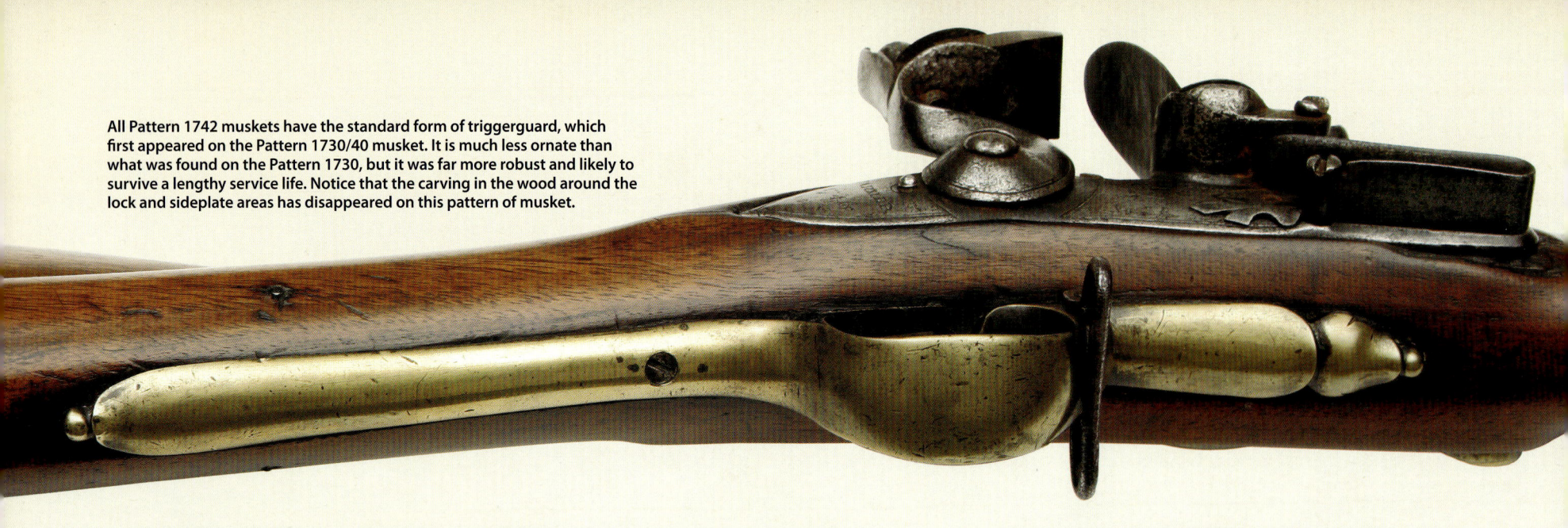

All Pattern 1742 muskets have the standard form of triggerguard, which first appeared on the Pattern 1730/40 musket. It is much less ornate than what was found on the Pattern 1730, but it was far more robust and likely to survive a lengthy service life. Notice that the carving in the wood around the lock and sideplate areas has disappeared on this pattern of musket.

Strikingly, more often than not these markings are actually American-applied tracking numbers. How do we know this? It is simple, really. First off, many of these markings are sloppily engraved, but that alone doesn't prove anything. The best indication is that these markings, taken as a group, simply do not relate to British army structure and do not match how the British applied their markings. Redcoat regiments used markings like "B/25" or "2/25." Both of these "rack numbers" mean the same thing – musket number 25 of the second most senior company in the regiment. The same goes for "H/4" and "8/4."

With that system in mind, what is one to make of wristplate markings like " Z/66" or "XXXI/43"? No British line regiments had anywhere near 26 or 31 companies, so these markings must be part of a distinctly different tracking system. Nothing bears this out like the "Gun Role" (or "account") of the arms possessed by Capt. Edmund Wells' Company of Connecticut soldiers during the spring of 1757, which recorded some truly curious fraction-style "rack numbers." Illustrated in Jim Mullins' excellent book *Of Sorts For Provincials* on p. 101, numerators are shown to have included Arabic numbers amongst Roman numerals and English letters, all atop denominators running between one and 100. Aside from these crazy American tracking fractions, sometimes the simple initials of long-dead owners are found inscribed into the stocks of Pattern 1742, at times indicating a succession of owners.

These same colonial owners, far removed from the Ordnance/army system of arms replacement and repair, had to make do in order to keep their muskets working. These repairs involved the incorporation of many non-standard parts, including those from the muskets of other nations as well as those made by local smiths. Therefore, an occasional home-spun replacement part should

Atypical unit markings like the one shown on this wrist plate are believed to be of mid-18th century American origin.

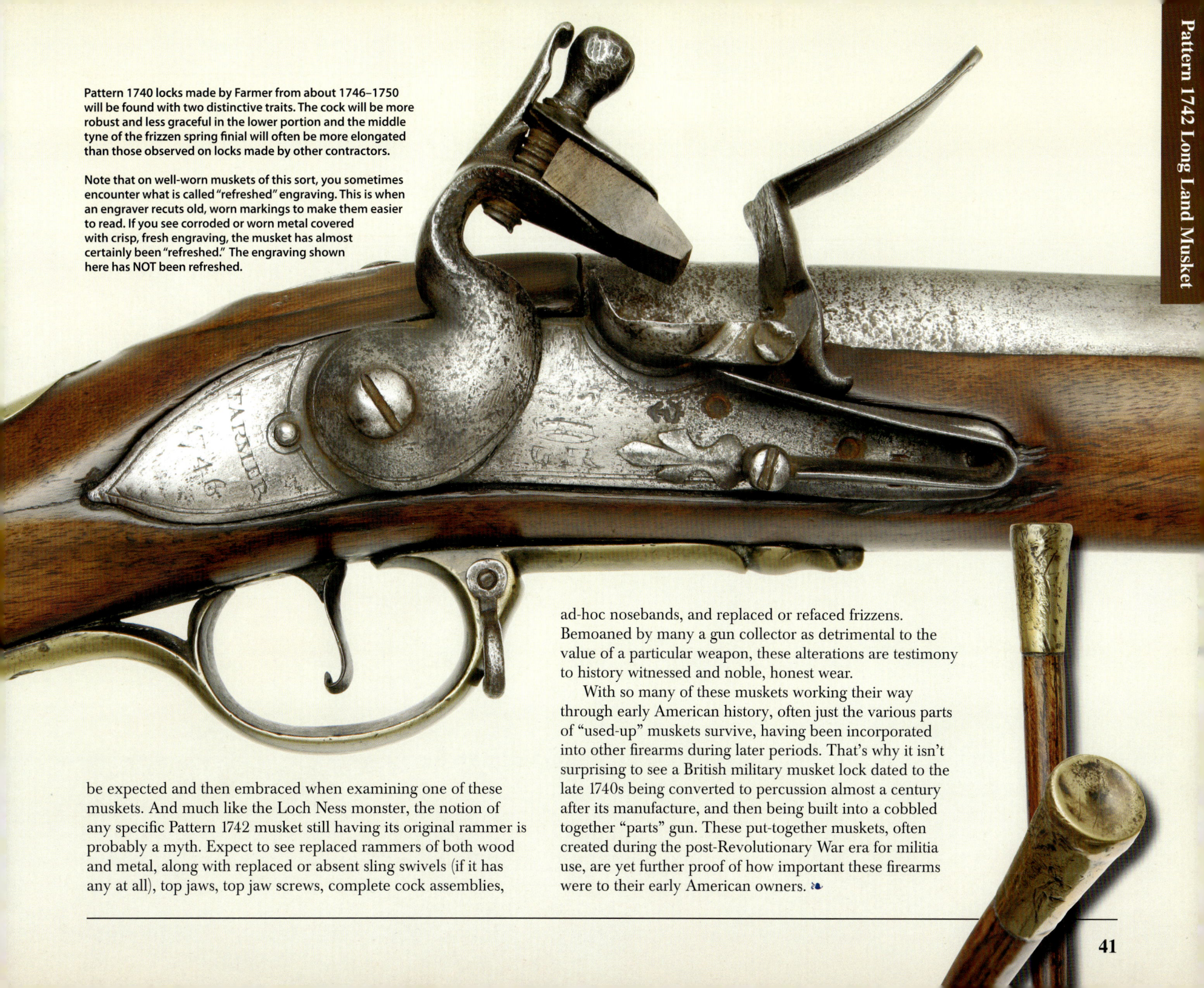

Pattern 1740 locks made by Farmer from about 1746–1750 will be found with two distinctive traits. The cock will be more robust and less graceful in the lower portion and the middle tyne of the frizzen spring finial will often be more elongated than those observed on locks made by other contractors.

Note that on well-worn muskets of this sort, you sometimes encounter what is called "refreshed" engraving. This is when an engraver recuts old, worn markings to make them easier to read. If you see corroded or worn metal covered with crisp, fresh engraving, the musket has almost certainly been "refreshed." The engraving shown here has NOT been refreshed.

be expected and then embraced when examining one of these muskets. And much like the Loch Ness monster, the notion of any specific Pattern 1742 musket still having its original rammer is probably a myth. Expect to see replaced rammers of both wood and metal, along with replaced or absent sling swivels (if it has any at all), top jaws, top jaw screws, complete cock assemblies, ad-hoc nosebands, and replaced or refaced frizzens. Bemoaned by many a gun collector as detrimental to the value of a particular weapon, these alterations are testimony to history witnessed and noble, honest wear.

With so many of these muskets working their way through early American history, often just the various parts of "used-up" muskets survive, having been incorporated into other firearms during later periods. That's why it isn't surprising to see a British military musket lock dated to the late 1740s being converted to percussion almost a century after its manufacture, and then being built into a cobbled together "parts" gun. These put-together muskets, often created during the post-Revolutionary War era for militia use, are yet further proof of how important these firearms were to their early American owners.

42

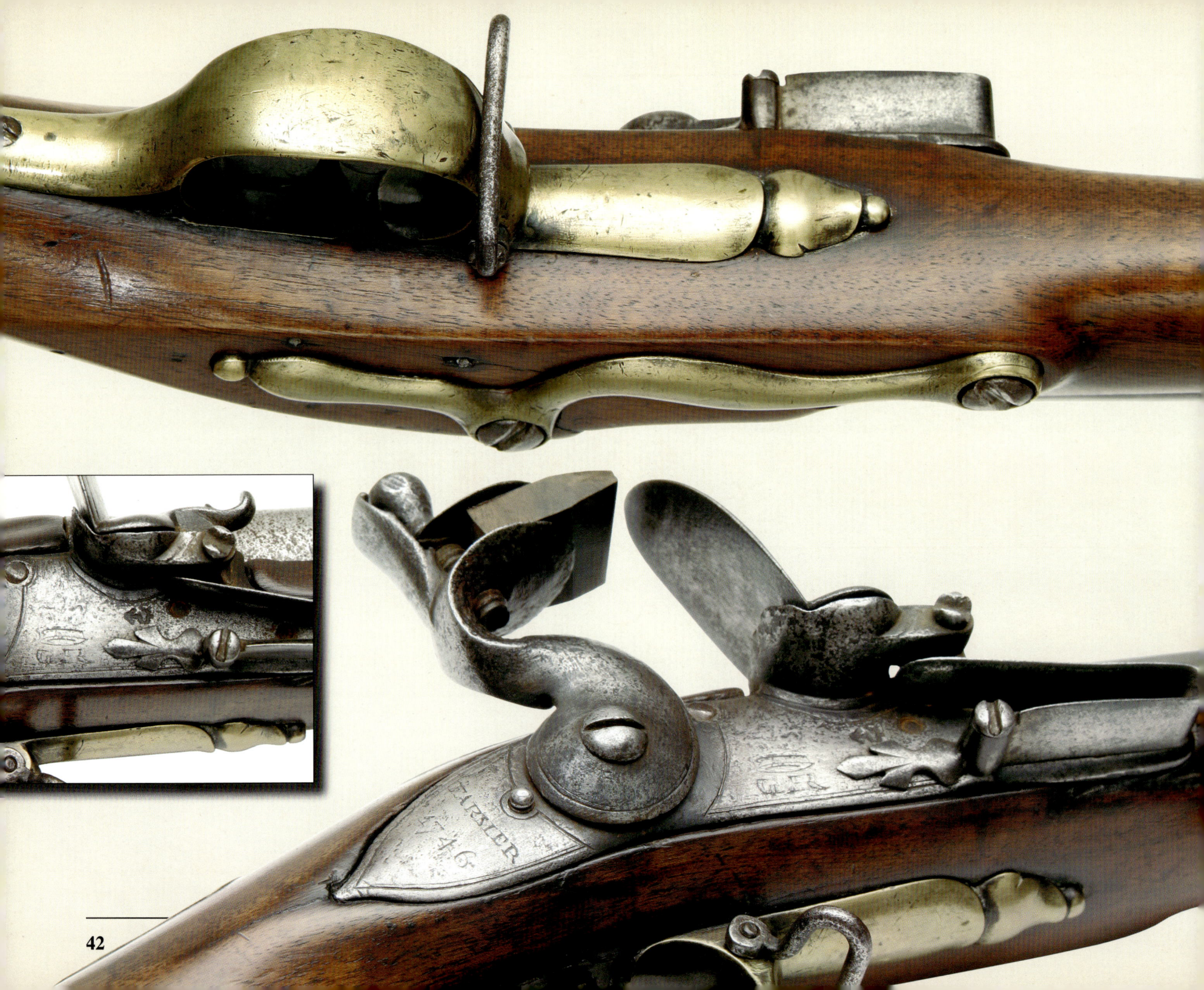

Pattern 1742 Long Land Musket

Ordnance storekeeper's mark.

Typical buttstock form for the Pattern 1742 muskets. The buttstock is heavier in cross section than what is found on the Pattern 1730.

44

Pattern 1742 Long Land Musket

Various views of the forestock, muzzle and furniture of a Pattern 1742 musket. Note that the top rammer pipe (with filed decoration) and copper noseband are American additions. The sling swivel shown here, heavier than that on the Pattern 1730, is the classic form for this pattern of musket.

45

TOWER
1742

GR

Pattern 1742 Long Land Musket

(above)
A wonderfully grimy lock mortise showing all the filth of decades of use. Note the musket's assembly number XII, which also appears on the edge of the lock's mainspring. The touch hole is wonderfully blown out, leaving gaping evidence of its lengthy service life.

(left page)
On this Pattern 1742 musket's lock, the cock, cock screw and mainspring screw are replacements. The "F" on the inside of the lockplate possibly stands for Farmer. The crowned number 2 is the mark of an Ordnance lock inspector.

(bottom two views)
The lock on another Pattern 1742 musket. Again we see the "F" mark on the inside of the lockplate, which may indicate the contractor Farmer. The fact that Farmer signed and dated this lock in 1746 supports this and may help to identify similarly marked "Tower" locks like on the previous page.

47

This page shows a number of views, including two examples of American "rack" numbers from Pattern 1742 muskets, a barrel marked to Ordnance contractor Edward Jordan, and a modern replacement sideplate marked on the underside by restoration supplier Reeves Goehring.

This view shows an original sideplate on a Pattern 1742 musket. Note the Thomas Hollier "H" mark and broad arrow on its underside, along with an "XII" musket assembly number. Note that the style of cast "H" and broad arrow used by Hollier are a different form than the ones illustrated in earlier chapters of this book.

Pattern 1742 Long Land Musket

49

THE PATTERN 1748 LONG LAND MUSKET

This rare, transitional musket is essentially a Pattern 1742 made with a steel rammer.

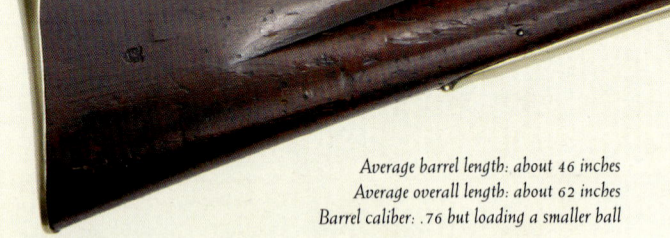

Rarity: Very Rare

Average barrel length: about 46 inches
Average overall length: about 62 inches
Barrel caliber: .76 but loading a smaller ball

Much like the so-called Pattern 1730/40 musket, "Pattern 1748" is not a proper Ordnance designation, but a term of convenience used by collectors to describe a transitional Land Service musket. The "Pattern 1748" is essentially a Pattern 1742 having some upgraded features, the most important of which is a steel rammer.

As new, improved components began arriving in Ordnance stores during 1748, they were sent out to the "setters up" and built into the British Ordnance's first standard musket to be made for a steel rammer. The Irish Ordnance, interestingly, was clearly ahead of the Tower in this instance, having procured steel-rammered muskets as early as 1724.

Continuing the trend of Bess' evolutionary yo-yoing weight fluctuation, she is once again getting thinner with the Pattern 1748. The inclusion of a smaller-diameter steel rammer removed the need for the heavier forestock seen on previous, wooden-rammered patterns. This reduction in rammer girth allowed the Ordnance's firearm architects to reduce not only the size of the rammer channel, but also the web of wood along the forestock separating the extremes of the rammer and barrel channels by approximately $\frac{1}{8}$". The new rammer also required smaller pipes than those used to secure a wooden one, so a reduction to the interior diameter of the newly designed upper pipes was also necessary.

Capping off this trim forestock is another British Ordnance first – a cast-brass nose cap – which became standard until well into the 19th century. Unlike the simple sheet-metal nosebands often added at Ordnance, regimental and other non-official levels, these cast caps are much heavier and completely enclose the end of the forestock, more effectively preventing the splitting encountered during hard service. This new innovation seems almost at odds with the archaic "banana"-shaped lockplate of the Pattern 1740 lock, which continued to be used on these muskets.

Firm	Supply Dates	Observed Dates*	Location
Farmer, James	1747-51	1747, 1750	Birmingham
Jordan, Edward	1747-50	1747	" "
Smith, John	1747-48	None	" "
Vaughan, John	1747	None	London
Wood, John	1748	1748	London
Tower (of London)	N/A	None	London

There are sure to be locks not recorded in this chart

Based upon how many times photos of Pattern 1748s have appeared in print during the last couple of decades, you might be led to believe that the Pattern 1748 is a fairly common musket. Not so! As of this writing, a grand total of nine specimens have been identified. And none of these muskets have seen much, if any, use. With the exception of some 20th century hanky-panky on a few locks, they are all in superb condition. Of these nine examples, eight are in museum collections, so the chance of encountering one for sale is slim to say the least.

All nine of these "unissued" Pattern 1748s are marked to the Second Battalion of the 23rd Regiment of Foot (also known as the Royal Welsh Fusiliers). The Second Battalion only existed from 1756 to 1758. Thanks to the meticulous record keeping of the Colonial

Production Period:
December 1748 to June 1749 (set up by contractors), then 1750–55 (set up at the Tower)

Total Production:
4,470 during the first period, possibly 1,000 during the second period

Lock Dates Observed:
1747, 1748 and 1750

Pattern 1748 Long Land Musket

Williamsburg Foundation (where seven of these muskets reside), it has been possible to trace the exact origins of this group.

All of these Pattern 1748s were sold to Colonial Williamsburg by W. Keith Neal in late 1950. Their origin was Flixton Hall, the Suffolk, England, ancestral home of Maj. Gen. Sir Allan Shafto Adair, commanding officer of the Guards' Armoured Division during WWII. Rather than being an extremely lucky Brown Bess collector, General Adair was the indirect descendant of William Adair, who had acquired the mansion and its muskets two centuries earlier. William Adair made a fortune during the 18th century as an "army agent." Army agents were business managers for particular regiments or officers, and as such they were the official channel through which muskets flowed between the Tower of London and the regiment being supplied.

In addition to being the agent of a number of other units and individuals, William

Pattern 1748 muskets continued to use the attractive and serviceable Pattern 1740 lock. This example was supplied by contractor Edward Jordan and is dated 1747. Since locks were stockpiled for later assembly under the Ordnance System, it should not surprise us that that date on this lock predates the date of the actual musket pattern.

It is worth noting that the Pattern 1748 musket used as illustrations for most of this chapter is in truly superb condition and gives an excellent impression of the quality that the Ordnance and its contractors were capable of delivering.

51

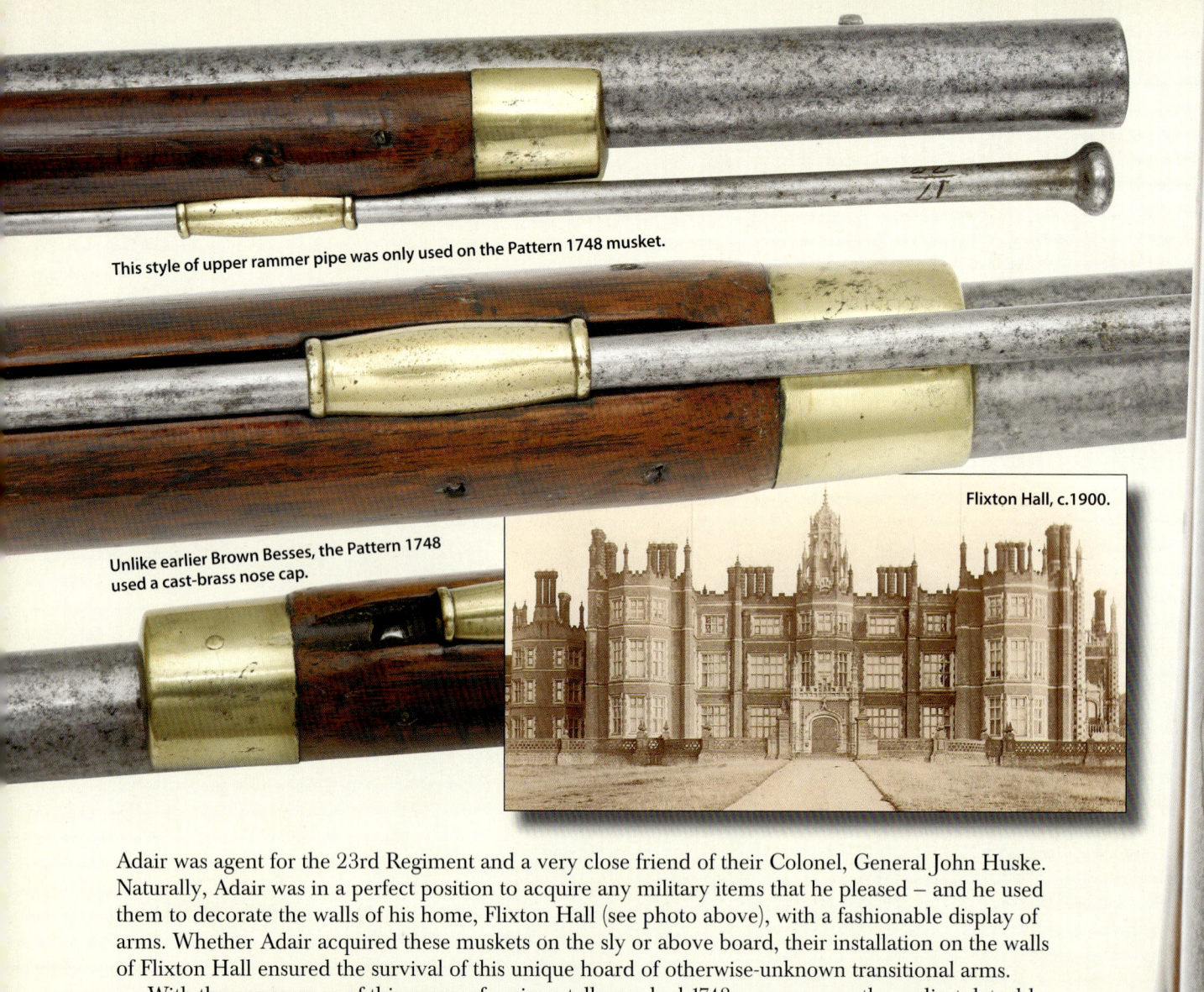

This style of upper rammer pipe was only used on the Pattern 1748 musket.

Unlike earlier Brown Besses, the Pattern 1748 used a cast-brass nose cap.

Flixton Hall, c.1900.

Adair was agent for the 23rd Regiment and a very close friend of their Colonel, General John Huske. Naturally, Adair was in a perfect position to acquire any military items that he pleased – and he used them to decorate the walls of his home, Flixton Hall (see photo above), with a fashionable display of arms. Whether Adair acquired these muskets on the sly or above board, their installation on the walls of Flixton Hall ensured the survival of this unique hoard of otherwise-unknown transitional arms.

With the appearance of this group of regimentally marked 1748s, we now see the earliest dateable evidence of the Board of Ordnance applying full versions of its "house style" of engraving. Although the inclusion of the ornate crest of the Royal Welsh Fusiliers is rightfully seen as exceptional, the fact that it is engraved along the top of the barrel, a few inches from the breech, shows adherence to the Ordnance's preferred sweet spot for the regimental indicator. In most cases, a simple regimental numeral followed by "REG^T" would have been engraved here. Wristplates were normally reserved for company and weapon number designation, engraved in the classic "rack number" or "fraction" format. The top "company" figure was either an Arabic numeral or a capital letter, while the individual musket number was a number under 100. Muskets had matching bayonets engraved with

52

Pattern 1748 Long Land Musket

(left)
A close-up view of the bayonet lug, dovetailed and brazed in place.

(bottom right)
With the first appearance of a steel ramrod in an Ordnance musket, there was now an opportunity to engrave "rack" numbers into this essential and easily misplaced component.

the same designation. Since the Pattern 1748 was the first British musket issued with a steel rammer, it created a need for this relatively-expensive piece of hardware to be included in the marking system. Within this group of nine Pattern 1748 muskets, we find rammers engraved with the same "rack numbers" found on their musket's wristplates and bayonets. Two still have matching bayonets and rammers while a number of other muskets have either one or the other from its original set, as originally put together at the Tower of London in 1756.

The threading on the bottom end of the ramrod was meant to secure a cleaning jag.

The year 1750 saw the last production of the Pattern 1740 lock with its beautiful, banana-shaped lockplate. Notice the musket assembly number on the sideplate screw shown below.

FARMER
1750

JORDAN
1747

54

Numerous views of the Pattern 1748 musket's stock and brasswork.

Ordnance storekeeper's mark

Classic Ordnance-style company and weapon number.

Radical notch

Pattern 1748 Long Land Musket

(left) Curiously, the barrel tang of this Pattern 1748 musket has been repaired with an exceptionally well executed brazing job. Note the fine, yellow line of brass that is clearly visible traversing the tang below the Ordnance inspector's mark. Certainly of the period, it is impossible to say whether this repair was done before or after it was built into this musket.

56

The touch hole and pan on this Pattern 1748 musket are in exquisite, pristine condition. The small diameter of the former and the sharp edges of the pan's depression indicate an almost complete lack of actual firing having been endured by this weapon. Also visible is the brazed repair to the barrel tang discussed on the previous page.

Pattern 1748 Long Land Musket

The "H" mark of Thomas Hollier, indicating a production date no later than 1754, the year he died.

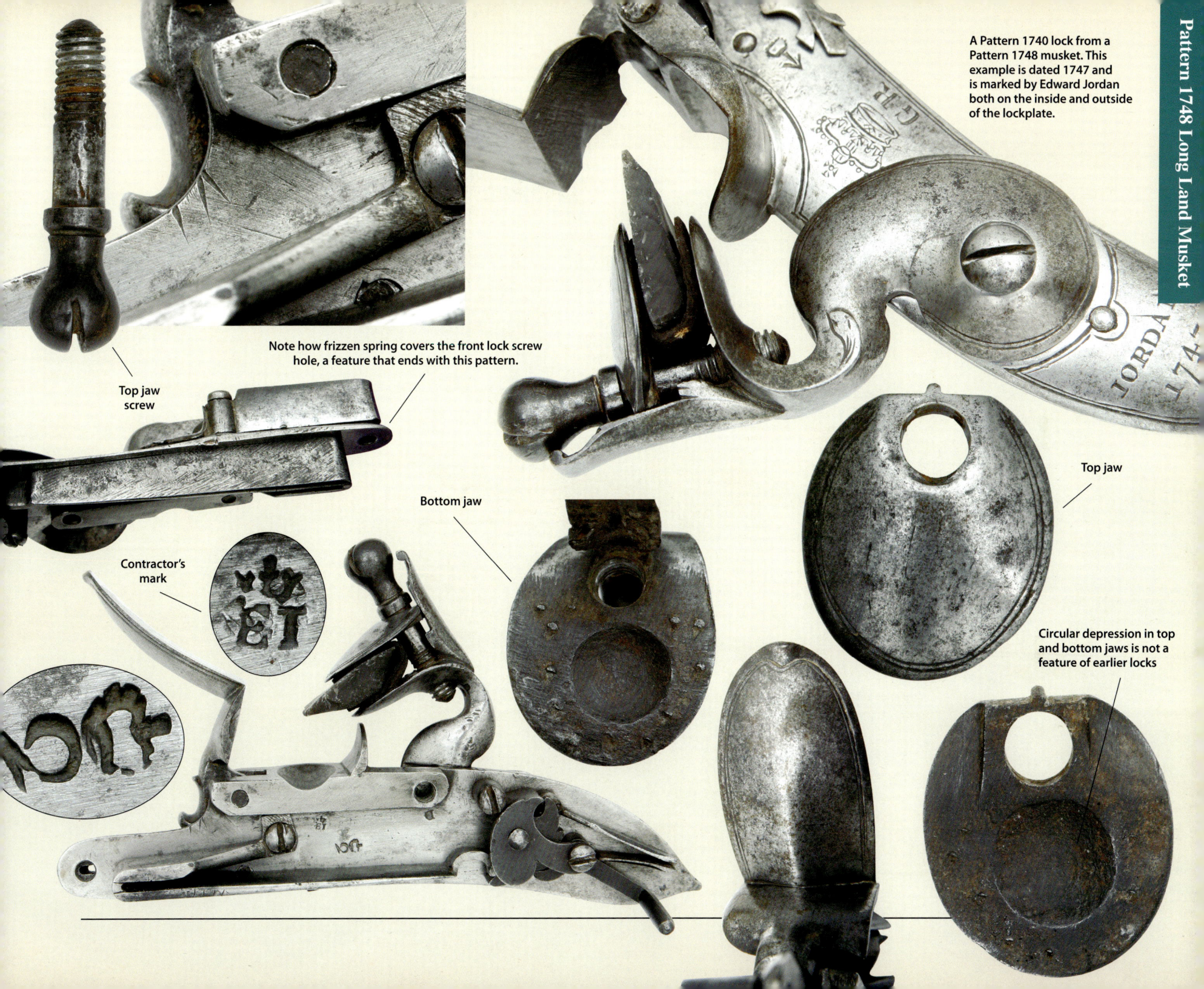

Pattern 1748 Long Land Musket

A Pattern 1740 lock from a Pattern 1748 musket. This example is dated 1747 and is marked by Edward Jordan both on the inside and outside of the lockplate.

Top jaw screw

Note how frizzen spring covers the front lock screw hole, a feature that ends with this pattern.

Contractor's mark

Bottom jaw

Top jaw

Circular depression in top and bottom jaws is not a feature of earlier locks *(right)* This crown mark struck into the inside of the triggerguard bow is found on all Long Land Pattern muskets going back to the Pattern 1730. Because this area of the triggerguard is protected from excessive polishing, this mark usually survives intact.

60

In addition to the wrist plate and the ramrod, the bayonet socket is also engraved with the company and weapon number.

Pattern 1748 Long Land Musket

THE PATTERN 1756 LONG LAND MUSKET (BRITISH)

The final pattern of Long Land Musket has a less-curved lockplate and a new rammer pipe.

Rarity: Scarce

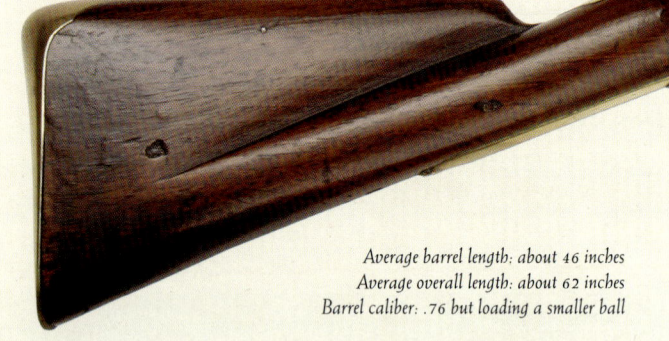

Average barrel length: about 46 inches
Average overall length: about 62 inches
Barrel caliber: .76 but loading a smaller ball

It is certain that all of the different Long Land pattern muskets are important in their own right, but one could argue that the Pattern 1756 really is the leader of the pack for some good reasons. Not because it's the prettiest or the highest in quality, but because this last incarnation of the 46-inch barreled Bess was produced in huge numbers – perhaps exceeding the combined totals of all previous patterns. The Pattern 1756 is simply "the" pattern of musket carried by Crown forces at the onset of the American Revolution.

Following the sensible, continuing evolution of the British martial musket, the Pattern 1756 exhibits a number of improvements. Retaining the rammer and cast-brass nose cap of the fleeting Pattern 1748, this musket improved upon the steel rammer system by introducing a long, trumpet-mouth pipe, which also serves as a useful, instant identifier. Anyone who has tried to return the rammer of an earlier pattern musket to its pipes during heated combat can tell you that this is a vast improvement! The principal is the same reasoning behind a simple funnel.

To the modern arms student, the curved or banana-shaped lockplate is an instant identifier, which screams "early Bess." Dispensing with this profile, the Pattern 1755 lock included with this musket has a less curved bottom to the lockplate's profile. While many references refer to the plate of the Pattern 1755 lock as "straight," it really isn't. If one removes the lock from one of these muskets and places a straightedge along the bottom of the lockplate, you can see how "unstraight" this edge really is. On a "banana" lockplate, the maximum gap between the lower edge and a straightedge is about 5–6 mm, while on a Pattern 1755 the gap is only 2–3 mm. Everything is relative!

Just as the shape of the lockplate is an indicator of vintage, so is the presence of an engraved date on the lockplate's tail end. The practice of dating lockplates was abolished after 1764, and all subsequent locks (and perhaps some reworked ones) were simply engraved TOWER behind the cock. Also disappearing at this time is the name of the manufacturing contractor from the outside of the lock. So early Pattern 1756 muskets will have locks engraved with dates and contractors' names, but later examples will not. Fret not, however, as these contractors still struck their mark (usually in some form of their initials), on the inside of the lockplates, so these makers can still be identified.

In a break from previous works, the authors of this book have conscientiously decided to

Firm	Observed Dates*	Location
Edge, Richard	1756-58, 1760, 1762	Wednesbury (outside Birmingham)
Farmer, James (& Galton)	1762	Birmingham
Galton, Samuel (& Farmer)	1756, 1762	" "
Grice, Joseph	1756, 1759-1762, 1764	" "
Haskins, George	1758	
Jordan, Edw. and/or Thos.	1756, 1759	" "
Perry, William	1756	
Stamps, Thomas	1756	
Vernon, George	1761, 1762	" "
Whately, John	1762	" "
Willetts, John	1762	Wednesbury (outside Birmingham)

There are sure to be locks not recorded in this chart

Production Period:
December 1756–1790
(but drastically reduced after 1768)

Total Production:
Between 200,000 and 250,000

Lock Dates Observed:
1756–1762, 1764

Pattern 1756 Long Land Musket (British)

make their discussions of the various musket patterns about the general type of musket, as opposed to celebrating a particular, spectacular specimen. This is why photos of different muskets are often mixed in together in an effort to illustrate best what the typical features of a musket pattern are. However, the musket that is used for most of the illustrations in this chapter (marked for the 23rd Regiment and engraved "5/94" on the wristplate) deserves special mention. Like so many of the surviving Pattern 1730s and all of the known Pattern 1748s, this musket is from Flixton Hall. As the best preserved musket in the lot, it is the example W. Keith Neal chose to keep for himself after purchasing the entire collection from the Adair family in 1950, and subsequently reselling everything else. It is probably the finest surviving Long Land musket of any pattern, at least as far as condition is concerned.

Having an essentially "as made" musket to study, disassemble, photograph and present to the readers of this work is especially valuable. From the superb quality of these photographs, one can see this arm as it would have appeared to its various makers and the soldier who would have been the original recipient of it. Original polish, finish and surfaces are present and exhibit the tell-tales of not only the way they were made, but the tools and materials used in the process. All of the crisp, new features of this particular musket should be kept in mind when studying a British Ordnance produced Land Pattern musket – because all of them looked like this when new!

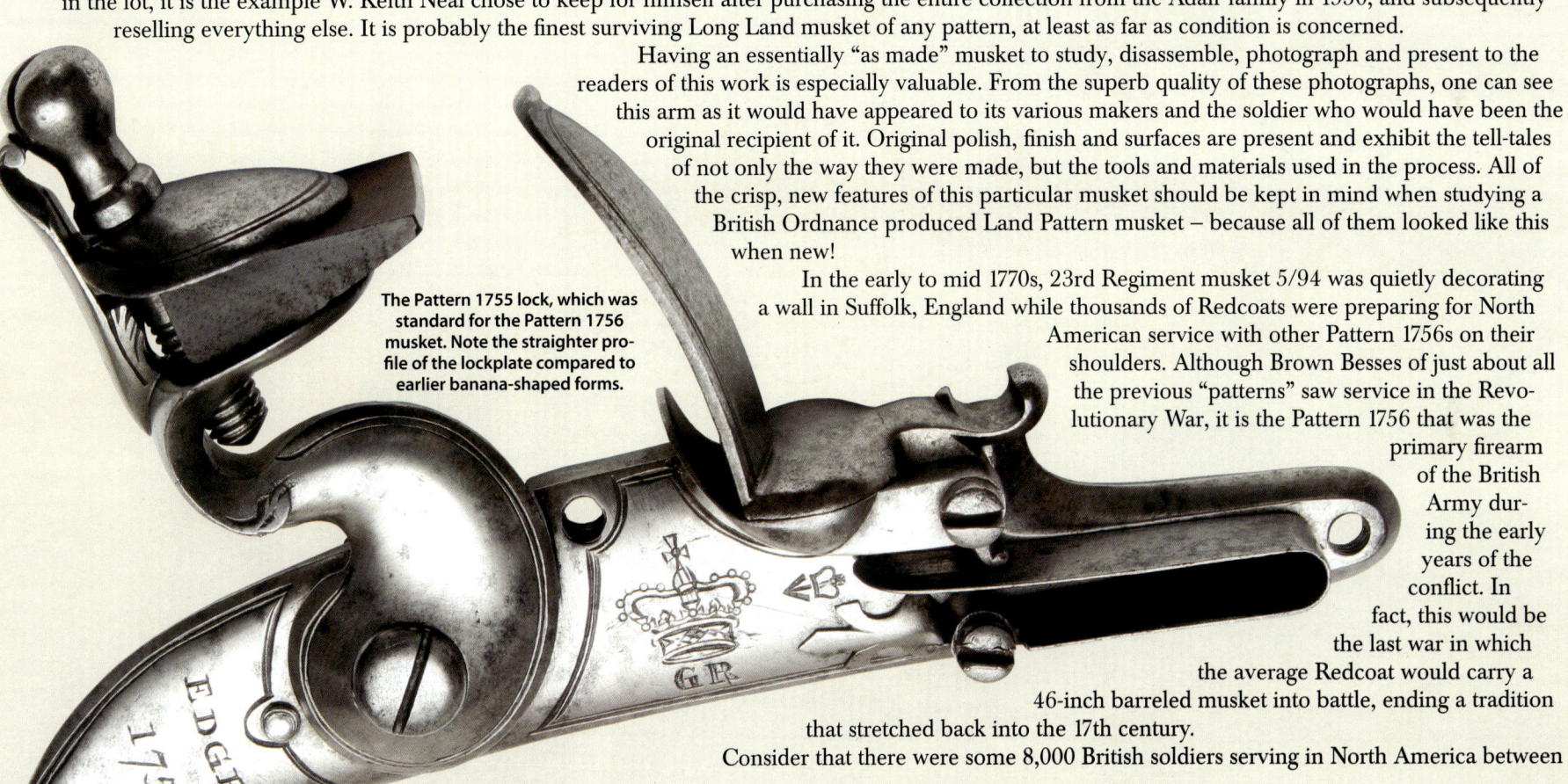

The Pattern 1755 lock, which was standard for the Pattern 1756 musket. Note the straighter profile of the lockplate compared to earlier banana-shaped forms.

In the early to mid 1770s, 23rd Regiment musket 5/94 was quietly decorating a wall in Suffolk, England while thousands of Redcoats were preparing for North American service with other Pattern 1756s on their shoulders. Although Brown Besses of just about all the previous "patterns" saw service in the Revolutionary War, it is the Pattern 1756 that was the primary firearm of the British Army during the early years of the conflict. In fact, this would be the last war in which the average Redcoat would carry a 46-inch barreled musket into battle, ending a tradition that stretched back into the 17th century.

Consider that there were some 8,000 British soldiers serving in North America between

63

Touch holes are not always absolutely dead center over the pan

Notice how the muzzle of the musket extends past the bayonet when properly fixed

The mark "FG" can be seen below on the breech of this Pattern 1756 musket. These initials stand for Farmer and Galton, a partnership in Birmingham that produced vast quantities of components for the Board of Ordnance.

Elongated sideplate flat typical of Pattern 1756 musket stocks

64

Pattern 1756 Long Land Musket (British)

Ordnance storekeeper's marks from this era are in the form of a crown over a pair of "GR" monograms, the right initials being a mirror image of those on the left.

Florida and the Canadian Maritimes (about 3,500 of which were stationed around Boston) during the Spring of 1775. It can, therefore, be assumed that a similar number of muskets were on hand, the vast majority of which would have been Pattern 1756s, either of English or Irish manufacture.

Regardless of the huge number of Pattern 1756s made, it seems that a comparable number of this and earlier patterns arrived in America during the wider time frame of 1737–1783. When confronted with the question as to why there are more surviving examples of the 1756, the answer must be simple – the 1730s, 1730/40s and 1742s got used up.

While relatively few of these earlier arms will be found to have full markings, it seems that with the Pattern 1756 the Board of Ordnance stepped up their practice of engraving regimental, company and weapon numbers into the arms before issue. Arms issued from the Tower more or less follow a fairly rigid format, and markings for better than a dozen different units that fought in America between 1775 and 1783 have been observed on Pattern 1756 muskets. To name but a few, there are those marked to the 4th "King's Own," 10th, 15th, 21st "Royal North British Fusiliers," 23rd "Royal Welsh

65

Fusiliers" (not from Flixton Hall), 31st and 43rd Regiments. Other units' Pattern 1756s are out there, and there are sure to be more waiting to come to light.

This final Long Land Pattern could never be described as "common," but it is the sole pattern that is eminently "gettable" for the collector or institution that wants one. So, depending on one's budget, one can afford to be a bit picky when choosing which Pattern 1756 to purchase. Further complicating the acquisition of a Pattern 1756 are the regimental markings described above. Original regimental markings add historical interest to a weapon and this affects price. The not-so-attractive by-product of this phenomenon is that some unscrupulous individuals have added spurious regimental markings to muskets that never had them in the first place...all in an attempt to get more money for muskets they have for sale. Given the amount of extra money involved, special attention should be paid to any regimental markings that are being offered as genuine.

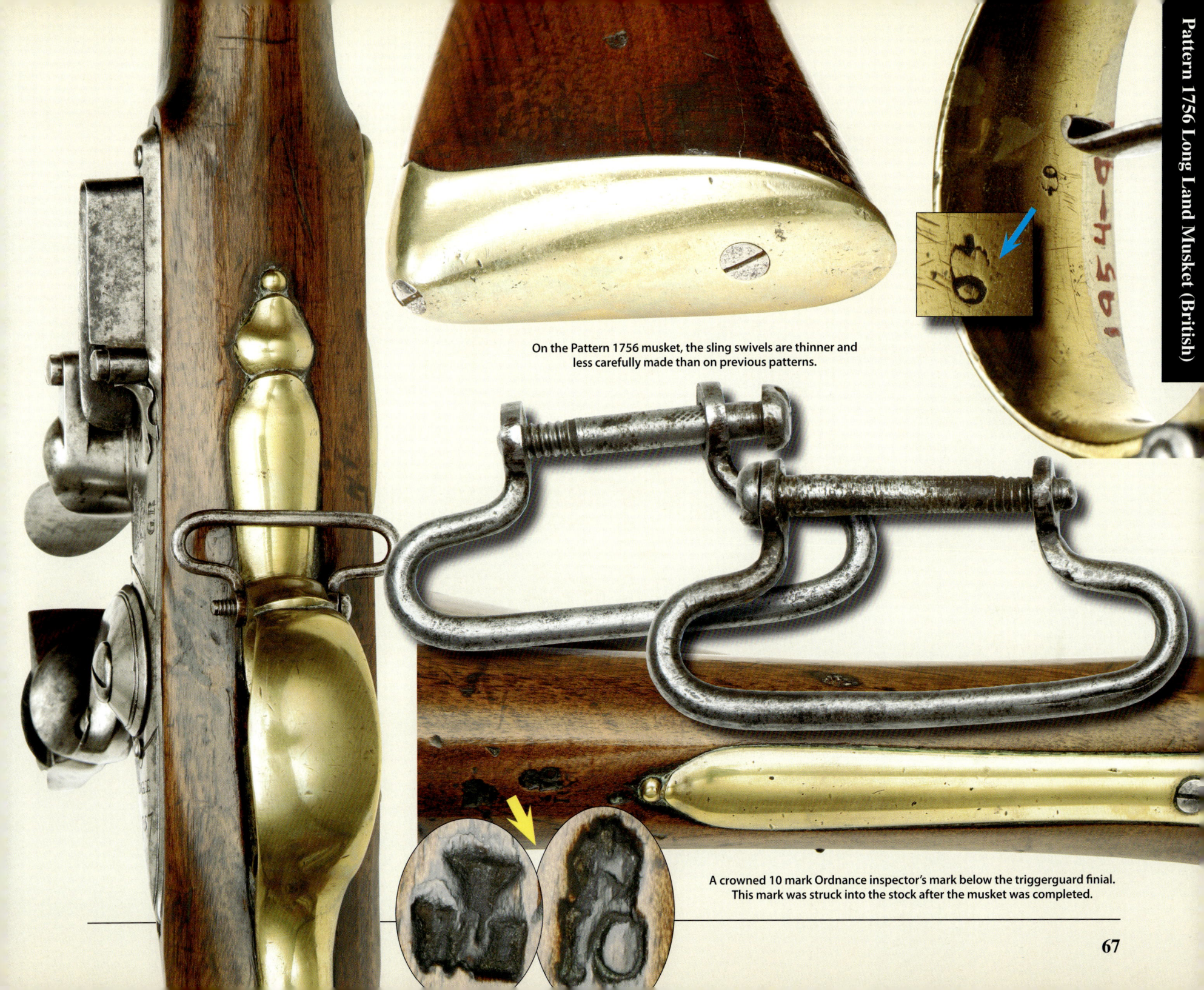

Pattern 1756 Long Land Musket (British)

On the Pattern 1756 musket, the sling swivels are thinner and less carefully made than on previous patterns.

A crowned 10 mark Ordnance inspector's mark below the triggerguard finial. This mark was struck into the stock after the musket was completed.

67

The initials of Farmer and Galton, the Tower barrel view and proof marks, as well as the crowned 5 Ordnance inspector's mark are all clearly shown in these views. The crowned 5 indicates that the barrel was properly vented, which explains its location next to the touch hole. This exact crowned 5 mark has also been observed on a number of Ordnance bayonets of the late 1750s. During this period, George Markby was the Ordnance inspector stationed in Birmingham and the crowned 5 may well be his mark.

Interior stock spaces were left unstained and unfinished on these muskets. However, the condition of most muskets today does not allow this to be observed. This musket, in almost new condition, gives us a rare opportunity to see what the inside of a Brown Bess really looked like when it was new.

Musket assembly number.

Pattern 1756 Long Land Musket (British)

Even Ordnance workmen slipped every now and then. Notice the scratch across the lockplate in the bottom right view, just under the crowned broad arrow government ownership mark.

Pattern 1756 muskets were equipped with Pattern 1755 locks. When new, the tumbler, bridle and sear achieved a color as part of their hardening process. See how they stand out from the other lock parts, which were polished afterwards to give them a brighter appearance.

Notice how tightly the frizzen covers the pan on a lock in this condition.

Notice the initials of lockmaker Richard Edge on the inside of this lockplate (see red arrow).

Pattern 1756 Long Land Musket (British)

Note the Pattern 1756's reduced stock swell in this area.

Improving on the steel ramrod system of the Pattern 1748 musket, the Pattern 1756 introduced a longer, trumpet-mouthed ramrod pipe. This made it much easier to put the ramrod back in the musket.

Pattern 1756 Long Land Musket (British)

73

THE PATTERN 1756 LONG LAND MUSKET (IRISH)

Irish-made muskets with several distinctive features.

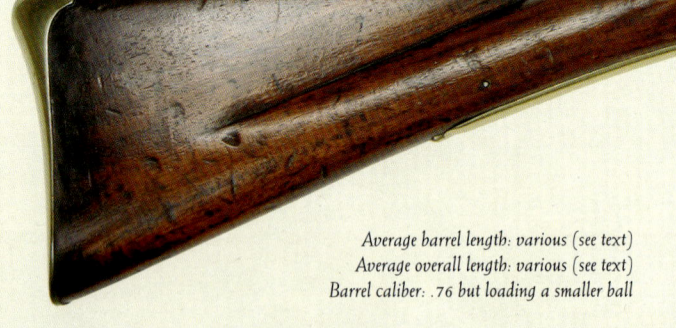

Rarity: Very Rare

Average barrel length: various (see text)
Average overall length: various (see text)
Barrel caliber: .76 but loading a smaller ball

Although it is little recognized today, arms students suffered a catastrophe in June of 1922. After being foolishly (or fiendishly) used as an ammunition dump during the Irish Civil War, a mysterious explosion with its ensuing fire destroyed most of the Irish Records Office's collection. Along with documents dating back 700 years, much of the records relating to both the British Army in Ireland and the materials produced for Dublin Castle were incinerated.

Established during the Commonwealth era of the previous century, Ireland had a separate military establishment, complete with its own Board of Ordnance tasked with supplying the British troops stationed on the isle. Initially occupied with the refurbishment of arms, it wasn't until the second decade of the 18th century that Dublin Castle began procuring truly Irish-made firearms. The earliest traceable musket is dated 1724 on the lockplate, and is strikingly similar to the forthcoming Pattern 1730, but was iron mounted and manufactured with a steel rammer and a noseband.

While there are numerous iron-mounted Dublin Castle muskets known from the 1720s and 1730s, few of the arsenal's products from the 1740s have been observed. One clearly Irish-made Pattern 1740 lock (marked "POWELL/1747") is known, though, suggesting that there was an Irish version (or perhaps versions) of the Pattern 1742 musket. However, all of these early Dublin Castle muskets must be considered great rarities, so our real story begins with the Pattern 1756 musket.

As it is generally accepted that things were done differently on the Emerald Isle, one expects early Irish products to differ somewhat from their English cousins. However, with a cut-and-dried Pattern 1756 musket being produced for the Tower of London, you might assume that Dublin Castle's version would (at the very least) be extremely close, especially after half a century of production under an "Ordnance system." But you would be wrong.

Maker*	Date of supply	Observed Dates
Alley, Lewis	after October 1770	
Collins, Matthew	before November 1771	
Dalton, Edward	after March 1776	
Dixon, John	after July 1760	
Govers, John		
Hutchinson, Michael	after December 1774	
Lord, Benjamin		
Lord, Francis		1762
Powell, William		1762
Ransford, James	after April 1759	
Scammon, John		
Thorpe, Thomas		
Trulock, James		1762
Trulock, Samuel	after 1766	
Trulock, Thomas		

*There are sure to be locks not recorded in this chart. Furthermore, it should be noted that not all of the above contractors supplying components to Dublin Castle were making musket locks. Those that have been confirmed are in bold.

74

Production Period:
c.1756–1775

Total Production:
Far less than 60,000 (based on a maximum output of 3,000 per annum)

Lock Dates Observed:
1760, 1762 (with identical locks in Short Land muskets dated 1767 & 1769)

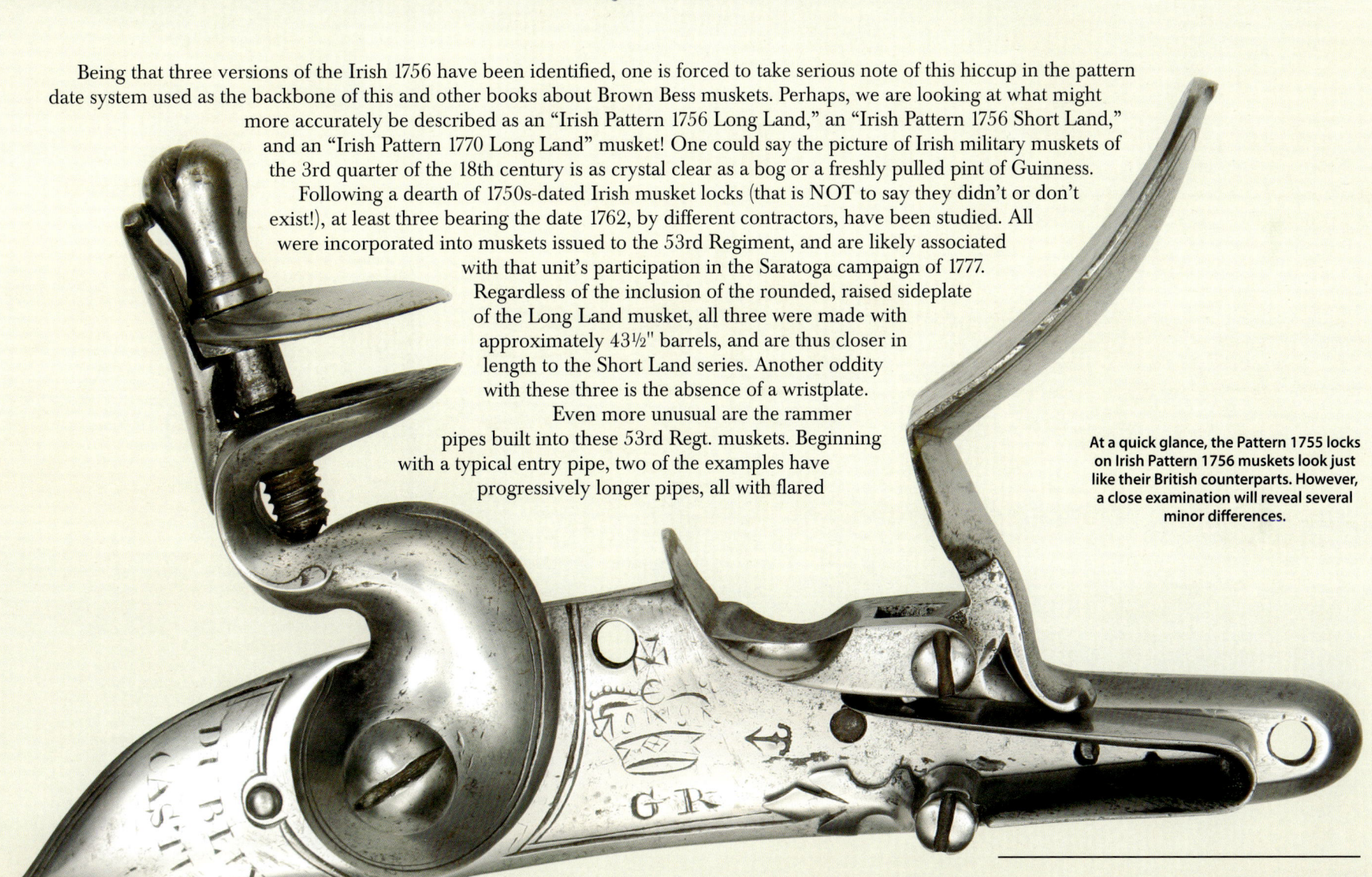

Pattern 1756 Long Land Musket (Irish)

Being that three versions of the Irish 1756 have been identified, one is forced to take serious note of this hiccup in the pattern date system used as the backbone of this and other books about Brown Bess muskets. Perhaps, we are looking at what might more accurately be described as an "Irish Pattern 1756 Long Land," an "Irish Pattern 1756 Short Land," and an "Irish Pattern 1770 Long Land" musket! One could say the picture of Irish military muskets of the 3rd quarter of the 18th century is as crystal clear as a bog or a freshly pulled pint of Guinness.

Following a dearth of 1750s-dated Irish musket locks (that is NOT to say they didn't or don't exist!), at least three bearing the date 1762, by different contractors, have been studied. All were incorporated into muskets issued to the 53rd Regiment, and are likely associated with that unit's participation in the Saratoga campaign of 1777. Regardless of the inclusion of the rounded, raised sideplate of the Long Land musket, all three were made with approximately 43½" barrels, and are thus closer in length to the Short Land series. Another oddity with these three is the absence of a wristplate. Even more unusual are the rammer pipes built into these 53rd Regt. muskets. Beginning with a typical entry pipe, two of the examples have progressively longer pipes, all with flared

At a quick glance, the Pattern 1755 locks on Irish Pattern 1756 muskets look just like their British counterparts. However, a close examination will reveal several minor differences.

75

mouths. While the upper pipes are similar to the British versions in profile, they were made without the raised molding at the end of the flared section. Being made of a more coppery-colored alloy than the rest of the brass mounts, it has been suggested that these oddball rammer pipes are American replacements. The third example has the usual style of pipes. Hmmm.

The next variation gets closer to the British Pattern 1756 Long Land. It was built with a 46" barrel and a wristplate, and since it includes an undated Dublin Castle lock, it may have been assembled later than the 53rd's muskets (Dublin Castle locks have been observed with locks dated as late as 1769). Marked for issue to the grenadier company of the 52nd Regt., it may be one of the "new set" of arms issued to them on Boston Common on the morning of June 17, 1775, the day of the Battle of Bunker Hill. As urgent times demand expediencies, this might be a sterling example of the Ordnance's issue practices turning topsy-turvy, with troops serving on the British Establishment (essentially the entire world, except for Ireland) receiving a new set of Irish arms. While the shorter arms of the 53rd Regt. were issued to battalion companies, perhaps the barrel on the 52nd Regt. musket is full Long Land length because it was meant for a grenadier company?

Linking this 52nd Regt. musket with those of the 53rd Regt. is a distinctive stock architecture marked by a drooping buttstock, a very much non-English feature. Additionally, this grenadier's musket has the same oddball rammer pipes as the 53rd's. Being that these components aren't easily lost and are almost impossible to break or wear out, the idea of them all being American replacements seems farfetched – especially on the arms of different regiments. It is therefore the opinion of the authors that they are distinctively Irish red herrings.

Last in this messy progression of Irish muskets is a lone example that perfectly conforms to the British Pattern 1756, and has survived as a complete "stand" with its numbered rammer and bayonet. Engraved on the top of the barrel is a large "24", indicating issue to the 24th Regt. While a lone numeral does not a regimental marking make, other Land Pattern muskets have been observed with this same "24" applied in an identical fashion, helping the case. Furthermore, company and weapon numbers engraved on the wristplate pretty much cinch the issue.

In all likelihood, this musket was issued to the 24th Regt. by one of the two warrants authorized at Dublin Castle in 1775. The first, dated April 21, decreed that 351 muskets and bayonets be delivered to the regiment as replacements for the worn out ones they had received (mostly) in 1766. To equip the new soldiers of the widespread Army augmentation of November 21, the Irish Ordnance issued another 170 stands. Not wanting a mix of older "long" and the brand new "short" arms in his regiment, the commanding officer of the 24th specifically requested Long Lands for these additional men. Wishing to oblige, Dublin Castle responded affirmatively, noting that it would be about a month until that number of Long Lands would be available for issue.

It is rare that a musket – let alone a complete stand – survives and can be illuminated further by some interesting and highly relevant documentation. But what else does this situation tell us? We know the Commanding Officer of the 24th Regt. (either Lt. Gen. Edward Cornwallis, the Colonel, or Lt. Col. Simon Fraser) was proactive in seeing to the uniformity of his unit. Therefore, it can be assumed that at the outset of the American Revolution, a significant percentage of less finicky-minded British regiments (who were armed with Irish arms) likely fell out for service with an array of muskets of different lengths, and indeed of different, yet-identified patterns. That is assuming that they can be accurately pigeonholed as "patterns" at all...

No wristplate

Irish Ordnance mark inside of the triggerguard

As a group, Dublin Castle muskets are far less consistent than British Ordnance examples. Before we move on to more pictures of what might be called a "standard" British Ordnance-style Irish musket, let's take a quick look at one of the interesting variations discussed in the text. When Gen. Burgoyne was proceeding down Upstate New York during the Saratoga Campaign of 1777, he left behind men from the 53rd Regiment of Foot at Fort Ticonderoga in order to secure his supply lines. Right before the first Battle of Saratoga, colonial troops attacked the fort and these men (and their muskets) were captured. The 53rd Regiment muskets have several interesting features. They have short barrels, measuring about 43.5 inches. They do not have wrist plates. The locks are shorter than typical examples. The furniture is all a bit unique in form — especially the middle two ramrod pipes, which are both flared. Regimental markings are found on the buttplate tangs, having the regimental number over the letter of the company and then the musket's "rack" number. The locks are all dated 1762 and bear the names of lock contractors — something that we do not often see on Dublin Castle muskets. The three lock contractors observed so far are "F. Lord", "Powell" and "I Trulock". Since these muskets were all presumably captured and fell into Colonial hands, they have rustic repairs and replacement parts. The most complete example is shown in this white box, which continues on the following page. The odd-ball features on these 53rd Regiment muskets have never been fully explained, although colorful theories abound.

Pattern 1756 Long Land Musket (Irish)

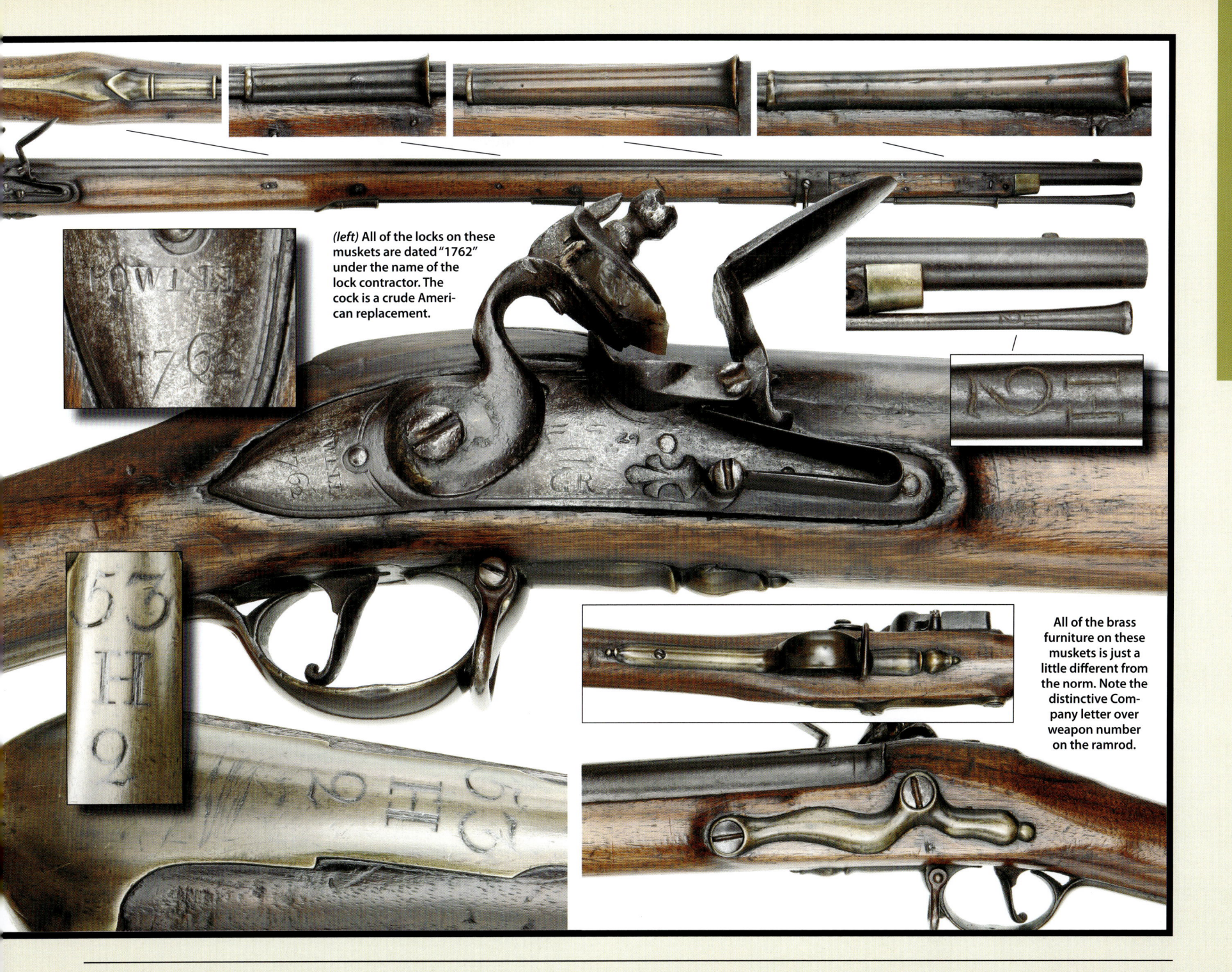

(left) All of the locks on these muskets are dated "1762" under the name of the lock contractor. The cock is a crude American replacement.

All of the brass furniture on these muskets is just a little different from the norm. Note the distinctive Company letter over weapon number on the ramrod.

This Irish ramrod pipe has a chunkier trumpet than British examples.

Barrel pin lug (green arrow).

The large "24" engraved into the barrel (see bottom right) indicates issue of this musket to the 24th Regiment.

Pattern 1756 Long Land Musket (Irish)

Crowned I mark, meaning unknown.

Sideplates on Irish Pattern 1756 muskets have a distinctive shape, being heavier and less graceful than the British versions.

Here, in this white box, we see a British Pattern 1756 sideplate for comparison with the Irish one shown above.

79

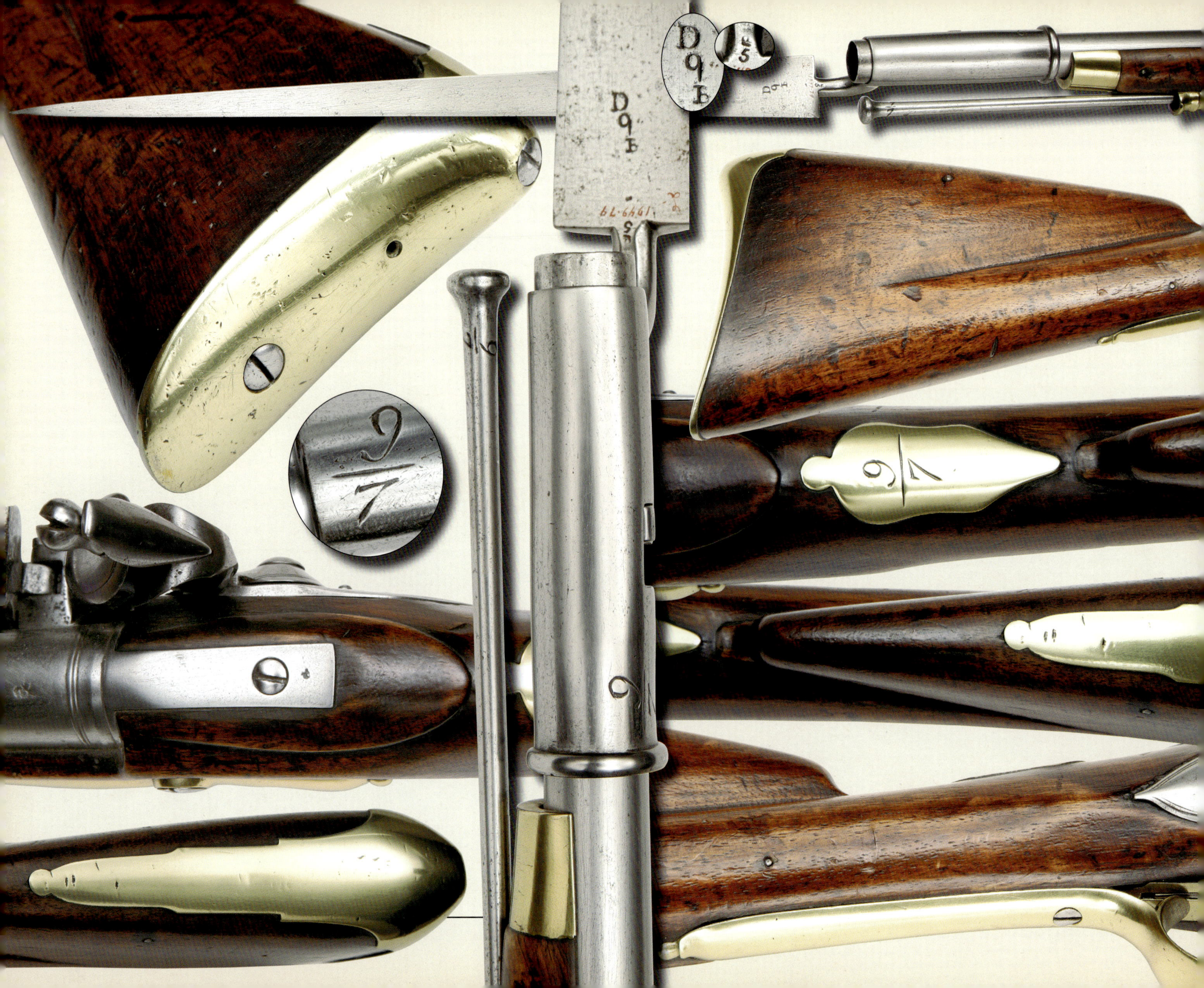
9/7

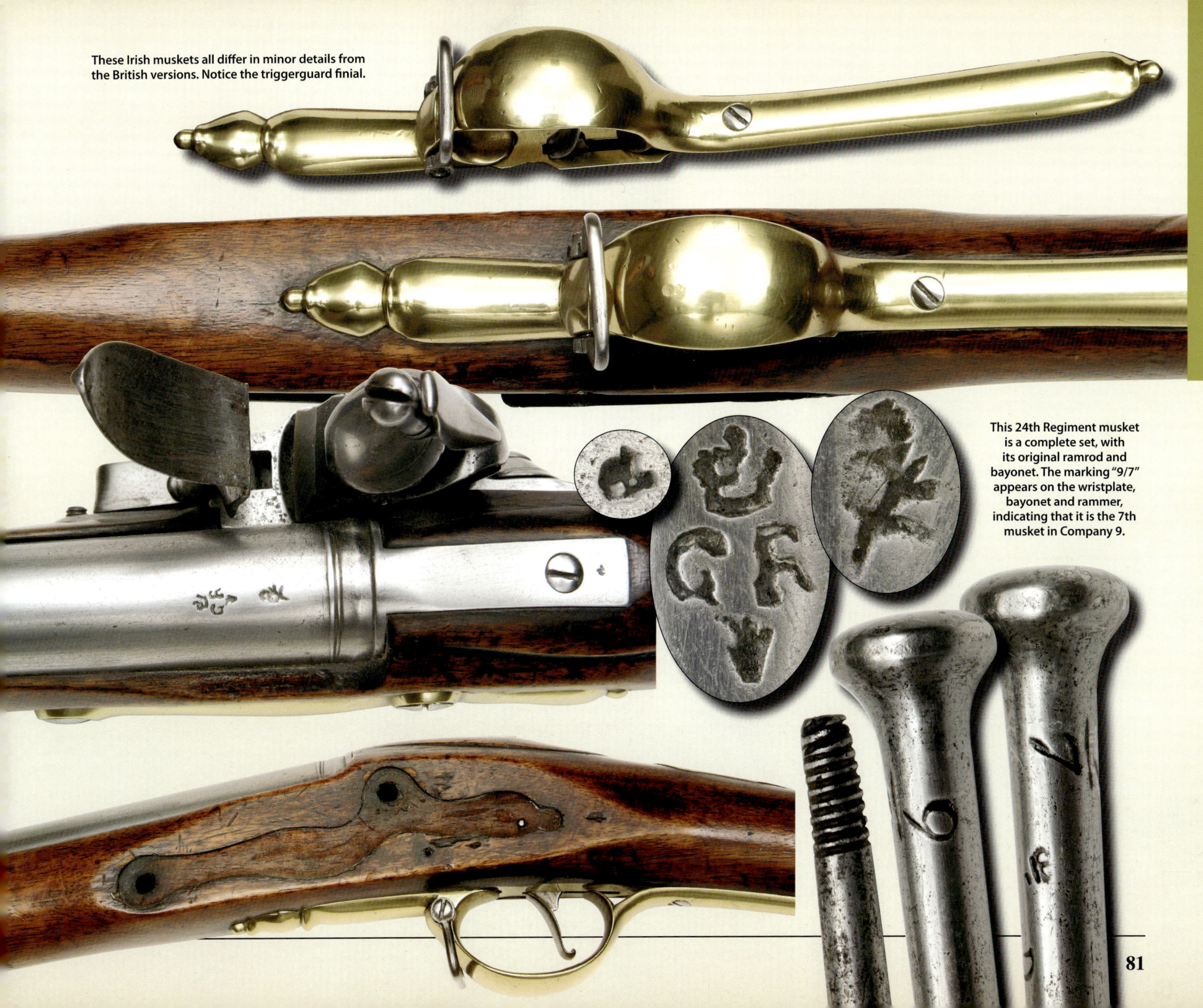

These Irish muskets all differ in minor details from the British versions. Notice the triggerguard finial.

Pattern 1756 Long Land Musket (Irish)

This 24th Regiment musket is a complete set, with its original ramrod and bayonet. The marking "9/7" appears on the wristplate, bayonet and rammer, indicating that it is the 7th musket in Company 9.

81

The frizzen of this musket shows an expertly executed repair. It has a dovetailed and brazed spline. A spline is a reenforcing piece of metal used in repair work and can be seen in the center of the frizzen where the face joins the pan cover. This was almost certainly a field repair done by a regimental armourer.

82

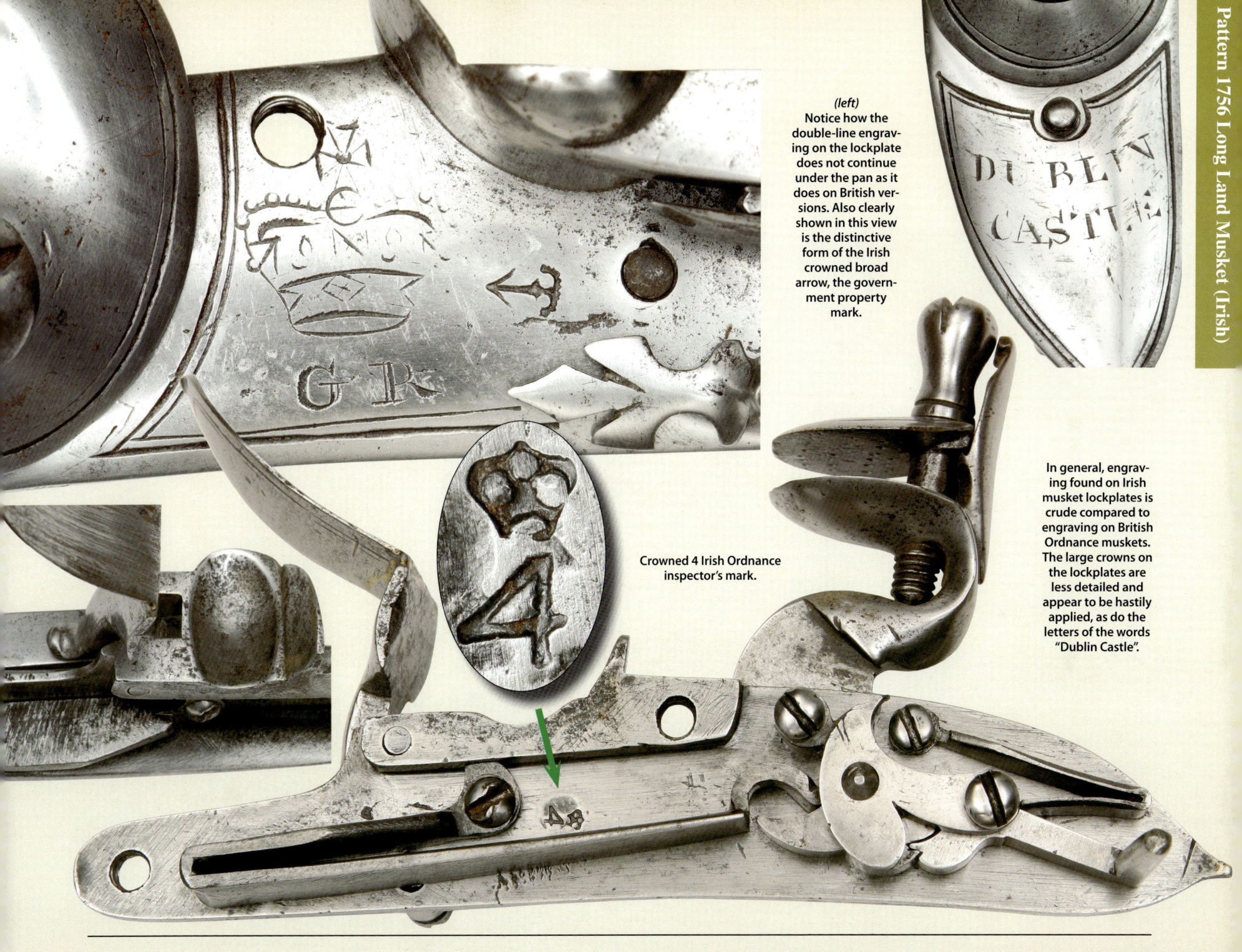

(left) Notice how the double-line engraving on the lockplate does not continue under the pan as it does on British versions. Also clearly shown in this view is the distinctive form of the Irish crowned broad arrow, the government property mark.

Crowned 4 Irish Ordnance inspector's mark.

In general, engraving found on Irish musket lockplates is crude compared to engraving on British Ordnance muskets. The large crowns on the lockplates are less detailed and appear to be hastily applied, as do the letters of the words "Dublin Castle".

Pattern 1756 Long Land Musket (Irish)

83

THE MILITIA OR MARINE MUSKETS
(PATTERNS 1757 AND 1759)

Rare muskets with unusual furniture.

Rarity: Rare

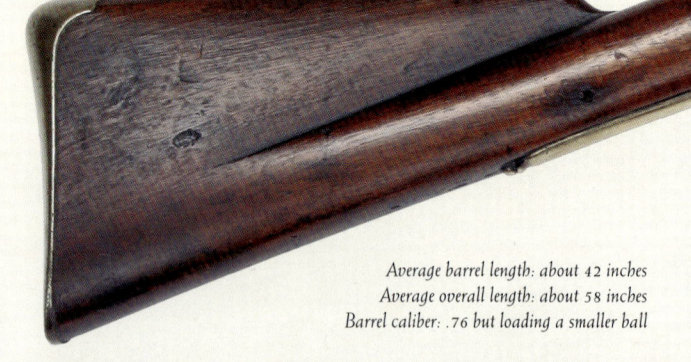

Average barrel length: about 42 inches
Average overall length: about 58 inches
Barrel caliber: .76 but loading a smaller ball

Developed in tandem with the Land Pattern musket series, albeit at a slower pace, was a succession of Sea Service muskets that are not covered in this book. Forgetting about workmanship and functionality, the arms intended for use aboard ship by grog-soaked and slops-clad sailors were certainly sturdy – but they were cheaply built. Bridging the gap between these two classes of muskets are those made specifically for issue to the Marines (soldiers serving aboard ship) and the various County Militias embodied after 1757 for various war efforts. Obviously, these two sorts of fighting men were deemed worthy of second rate arms by the Board of Ordnance.

Up until this point, our technological progression has dealt with a steady stream of improvements salted with the occasional bit of economy here and there. With the introduction of the "Militia or Marine" muskets during the Seven Years' War, we see the first appearance of a Land Service Pattern longarm designed specifically to save money.

Where could costs be cut without compromising durability and reliability? Certainly, the heart of the gun – its lock – needed to be of the highest quality produced by the Ordnance's contractors in Birmingham. So we don't see any reduction in quality there. The normal Pattern 1755 lock found on other Land Service muskets was employed. The musket's backbone – its barrel – was reduced from 46" to 42", which provided a small savings on labor and material. But the way they really saved money was on the musket's hardware, which was either cheapened or omitted.

Instead of the newly introduced steel rammer, the Pattern 1757 was only equipped with a wooden-rammered system even more archaic than that installed on the Pattern 1730 musket (it completely omitted the entry pipe). Harkening back to the days of the doglock musket, the lack of this pipe was also a trait found on mid-18th Century Sea Service muskets. Smartening up quickly, two years later the Ordnance revamped the pattern and brought it more in line with the contemporary Land Service musket, the Pattern 1756. Thus, the Pattern 1759 Militia or Marine musket has the same rammer pipe and steel rod setup as the Pattern 1756 Long Land musket, including the cast-brass nosecap at the end of the forestock. This upgrade mitigated some of the savings the Board had reaped with the Pattern 1757, with function obviously trumping economy. Well, sort of. The production period of the two patterns overlap for a period of almost 4 years.

The Pattern 1757 also saw the first appearance of the flat sideplate destined to become standard on later Short Land infantry muskets. Borrowing the deeply inlet flush-profile of the 3-screw Sea Service sideplates of previous decades, this new mount retained the contour

Distinguishing traits:	42" barrel, no wristplate, flat sideplate, 3-screw short-tang butt plate, plus...
Pattern 1757	*Wooden rammer held by 3 straight pipes (no entry pipe)*
Pattern 1759	*Steel rammer with entry pipe, long trumpet pipe & nosecap*
Lock contractors & dates:	Same as the Pattern 1756 Long Land musket, since these two patterns were all built with Pattern 1755 locks.

Militia or Marine Muskets

Production Period:
Pattern 1757: 1757–1764
Pattern 1759: January 1760–December 1776

Total Production:
At least 120,337 total. Pattern 1757, 71,323
Pattern 1759, at least 49,014

of the Land Pattern sideplate and was incorporated into the Militia or Marine muskets as well as their contemporary Sea Service muskets.

While we all love wristplates, this feature offered little to the 18th-century military man other than a platform on which to engrave various markings. One could argue that it added strength to the weakest part of the stock, but one could argue effectively against that point, too. Either way, it was completely omitted on both patterns of Militia or Marine musket.

So where, then, would unit markings be placed? Very often, markings are observed on the shortened tang of a new pattern buttplate created specifically for these arms. While the Short Land Pattern buttplate would also have a short tang, the Militia or Marine version was secured by three wood screws instead of the more normal two wood screws and traverse pin. This highly visible screw head coming through the tang of the buttplate is a key identifying feature of the Militia or Marine muskets.

With more than 120,000 made, it's a bit surprising that more of these muskets haven't survived. In fact, most specimens seen today will be found with buttplate markings reading something like "ABD" over "3/26", indicating issue to the Aberdeenshire Militia. This Scottish unit was first embodied in 1798, and most, if not all, of their surviving muskets are Pattern 1757s. With 40-year-old, wooden-rammered muskets being issued to a new unit during the period of a great Napoleonic threat, one gets an interesting picture of how the Ordnance system of supply and issue could work when pushed to extremes. The Aberdeenshire longarms have survived in relatively unworn condition, with at least one specimen still united with its numbered bayonet.

Other Militia or Marine muskets have been observed with markings applied in a more "standard" Ordnance format; for example, one has "M - BUCKS" (Buckinghamshire Militia) engraved on the barrel and "5/8" engraved on the butt plate tang. To date, only one Militia or Marine musket, of the 1759 pattern, has been noted with markings indicating issue to a Marine unit. With the company and weapon number of "15/79" on the buttplate tang, its triggerguard is engraved "PLYMo DN" on

The musket shown here is a Pattern 1757 that has been equipped with a steel ramrod.

Pattern 1755 lock, just like we see on the Pattern 1756 Land Service musket.

85

Note that this Ordnance storekeeper's mark has a broad arrow below the king's cypher. This broad arrow is not always found on other storekeeper's marks from this period.

Specific to the Militia or Marine musket is this distinctive buttplate tang secured by a wood screw as opposed to a lug and pin.

The markings on this buttplate indicate service with the Aberdeenshire Militia after 1798.

the long tang of the triggerguard, indicating its onetime possession by the Plymouth Division of Marines.

Considering the huge expansions of the county militias during the last half of the 18th century (by more than 30,000 men in 1757 alone) and the relatively small number of British Marines raised, it would seem that the vast majority of these muskets were destined for service — and eventual destruction — within the British Isles. Thus, few were likely used in North America.

The oval sling swivel made of wire is a post-issue replacement on this musket. When new, this musket would have had a sling swivel similar to that on a Pattern 1756 musket as shown on page 67.

Militia or Marine Muskets

The lack of a tail pipe is an almost instant way to identify one of the Pattern 1757 Militia or Marine muskets.

88

Note the lack of a nosecap, which tells us that this is the Pattern 1757 version of this musket. Also notice the larger diameter pipes and larger diameter rammer channel, other indicators of the Pattern 1757.

Militia or Marine Muskets

Note the original condition of this touch hole and lock mortise. The musket assembly number XXII (for twenty-two) also appears at the bend of the mainspring on the interior of the lock (see next page).

90

Militia or Marine Muskets

Views from the inside of the Militia or Marine musket's lock. Detached from its musket, there is no way to tell that this lock does not belong to a Pattern 1756 musket. The locks are identical, and in fact, the Militia or Marine locks were simply pulled from existing stockpiles of locks that would otherwise have been used to assemble the much more common Pattern 1756 Land Service musket.

Although made and marked by William Grice (see the "WG" initials), this lock was also marked by the same Ordnance inspector (crowned 2) who marked the Pattern 1755 lock made by Richard Edge shown on page 71.

THE PATTERN 1769 SHORT LAND MUSKET (BRITISH)

A shorter barrel length creates a new generation of infantry musket.

Rarity: Scarce

Average barrel length: about 42 inches
Average overall length: about 58 inches
Barrel caliber: .76 but loading a smaller ball

Breaking with decades of tradition, the Board of Ordnance decided to officially crop the barrel length of the standard infantry musket from 46" to 42" in the late spring of 1768. It was felt that the old Long Lands were too long and too heavy. This handy, new "Short" Land Pattern musket was assembled around the same shorter barrel built into Dragoon and Militia or Marine muskets for decades. Ignition was supplied by a Pattern 1755 lock engraved "TOWER" behind the cock. A newly designed short-tang buttplate, similar in profile to that of the Militia or Marine musket, was instituted.

So far, all of this sounds fairly straight forward. But when we get to the sideplate, things become complicated. During recent years, there has been some disagreement about what types of sideplates were used on the earliest of these Short Land Pattern muskets. This is an especially important topic, because these muskets played an important role in the American Revolution, where they were the newest and most modern muskets available to the British Infantry.

Some scholars believe that early Pattern 1769 Short Land Pattern muskets all had rounded-profile sideplates identical to those found on Long Land Pattern muskets. Then, in late 1775, the pattern was changed to incorporate a sideplate similar to that found on the Militia or Marine musket, which has a flat profile and was fully inlet into the stock. These scholars point to documentary evidence suggesting that this might be what happened. However, other scholars disagree, claiming that Militia or Marine-style flat sideplates were used on Pattern 1769s from the very beginning. They propose that the early type with a Long Land Pattern-style sideplate was either made in very small numbers or not at all. Disregarding the documentary evidence, these scholars point to physical evidence, observing that they have never seen a Pattern 1769 Short Land Pattern musket of the supposed early type having a Long Land Pattern sideplate – even on muskets known to have been captured very early in the Revolutionary War.

Throughout this book, the authors have tried to present the Brown Bess from various and differing perspectives, with no two "Patterns" discussed in exactly the same fashion. This chapter will analyze the scholarly debate about Pattern 1769 sideplates, examining the fascinating divergence that can take place when the interpretation of archival records seems almost totally at odds with the actual muskets that survive today.

Let's start with the documentary evidence. It is indisputable that the Pattern 1769 was approved in 1768 by both the Board of Ordnance and the King. Between this time and late 1775, a flood of components flowed into Ordnance stores; these were set up into muskets, which were readied for issue to the British Army. By 21 November 1775, some 37,000 muskets had been rough stocked. According to the first school of thought mentioned above, all of these muskets would have been made with the rounded-profile sideplate found on the Long Land series muskets and should be called the true "Pattern 1769". After that date, they say that the well-known, flat-surface sideplate began flowing into stores for inclusion on Land Pattern muskets, giving birth to what De Witt Bailey (the renowned researcher who first published this theory) terms the "Pattern 1769/75" musket.

Surviving bills for musket hardware dating from late November 1775 until 1777 (when the next musket pattern change took place) suggest that a maximum of 31,000 of the so-called

Production Period:
c.1768–1777

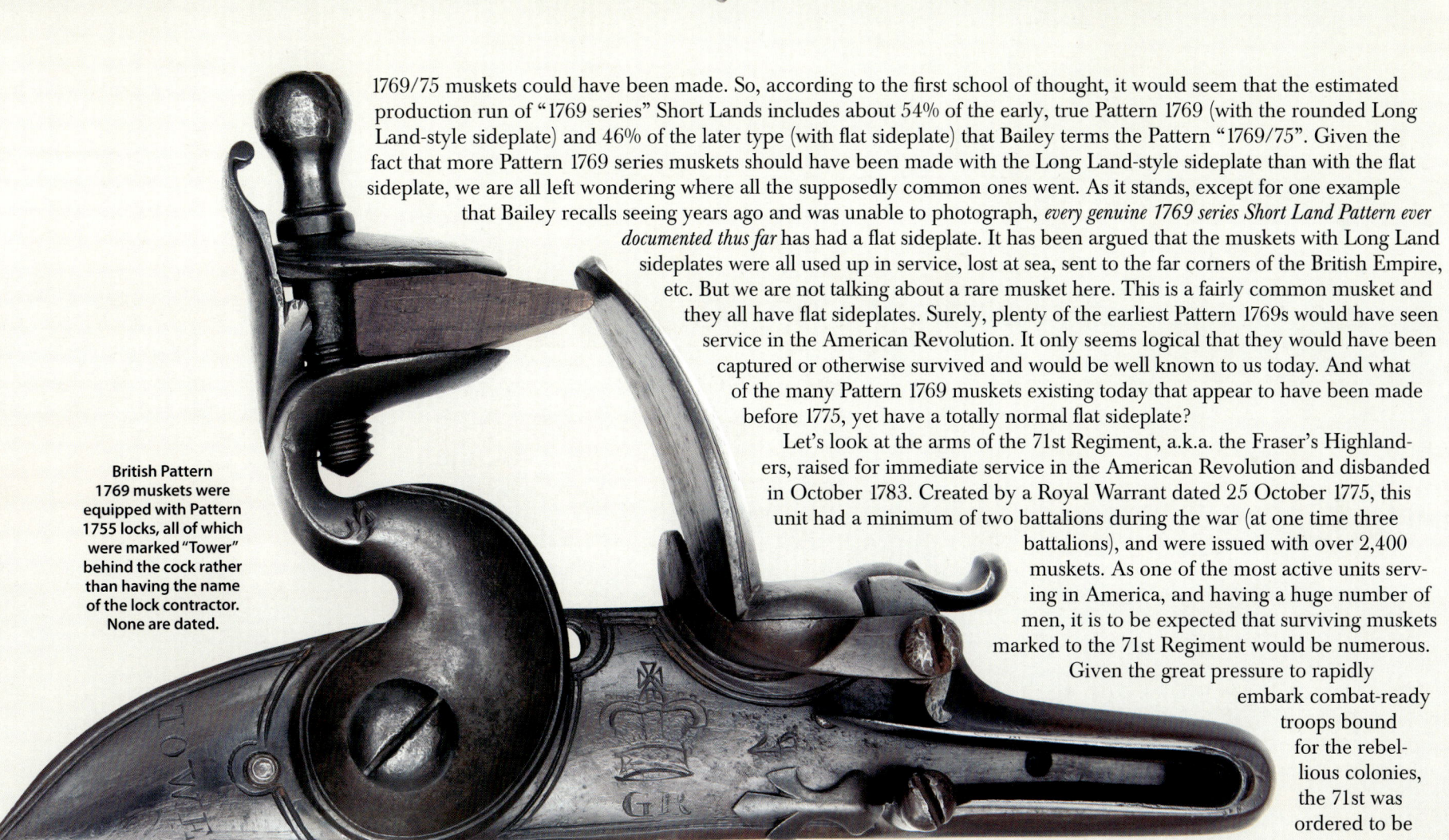

Total Production:
Less than 68,000

British Pattern 1769 muskets were equipped with Pattern 1755 locks, all of which were marked "Tower" behind the cock rather than having the name of the lock contractor. None are dated.

1769/75 muskets could have been made. So, according to the first school of thought, it would seem that the estimated production run of "1769 series" Short Lands includes about 54% of the early, true Pattern 1769 (with the rounded Long Land-style sideplate) and 46% of the later type (with flat sideplate) that Bailey terms the Pattern "1769/75". Given the fact that more Pattern 1769 series muskets should have been made with the Long Land-style sideplate than with the flat sideplate, we are all left wondering where all the supposedly common ones went. As it stands, except for one example that Bailey recalls seeing years ago and was unable to photograph, *every genuine 1769 series Short Land Pattern ever documented thus far* has had a flat sideplate. It has been argued that the muskets with Long Land sideplates were all used up in service, lost at sea, sent to the far corners of the British Empire, etc. But we are not talking about a rare musket here. This is a fairly common musket and they all have flat sideplates. Surely, plenty of the earliest Pattern 1769s would have seen service in the American Revolution. It only seems logical that they would have been captured or otherwise survived and would be well known to us today. And what of the many Pattern 1769 muskets existing today that appear to have been made before 1775, yet have a totally normal flat sideplate?

Let's look at the arms of the 71st Regiment, a.k.a. the Fraser's Highlanders, raised for immediate service in the American Revolution and disbanded in October 1783. Created by a Royal Warrant dated 25 October 1775, this unit had a minimum of two battalions during the war (at one time three battalions), and were issued with over 2,400 muskets. As one of the most active units serving in America, and having a huge number of men, it is to be expected that surviving muskets marked to the 71st Regiment would be numerous. Given the great pressure to rapidly embark combat-ready troops bound for the rebellious colonies, the 71st was ordered to be

Pattern 1769 Short Land Musket (British)

93

These views clearly show the "Tower" mark behind the cock, the large crown over "GR" and the crowned broad arrow government ownership mark.

equipped with 2,000 firelocks and bayonets out of Tower stores in early December 1775. Shortly thereafter, with their ranks filled, the 71st departed Scotland, with their new muskets in hand, on 21 April 1776. Of the seven 71st Regt. muskets studied, four are marked to the 1st battalion and three are marked to the 2nd battalion. A combined total of six of the muskets are classic Short Lands with flat sideplates and undated Pattern 1756 "Tower"-marked locks. The seventh purports to be a Pattern 1756 Long Land, but is in wretched, incomplete and compromised condition.

If early Pattern 1769 muskets had Long Land-style sideplates, we really should be seeing them on these guns. It seems highly unlikely that the arms issued to the 71st Regiment by an order of December 1775 were of a pattern so new to Ordnance stores that the sideplates had just arrived at the Tower a couple of weeks before. For that to be the case, these components would have had to have been immediately built into at least 2,000 complete muskets in time to be in the 71st's hands less than 4 months later, hundreds of miles to the north in Scotland. Furthermore, even if this were possible, it seems almost unbelievable that the Ordnance would break with tradition and issue brand-spanking new arms – that literally just walked in the castle door – to a regiment of raw Scottish recruits bound for colonial service.

Upon reflection, we must agree with the proposition that all, or nearly all, Pattern 1769 series Short Land muskets procured by the Ordnance were made with flat sideplates and that the early type with a rounded Long Land-style sideplate either never existed or was manufactured in very small numbers. We should remember that the Militia and Marine muskets initially made during the Seven Years' War had a flat sideplate of the exact Short Land type. So the sideplate first entered Ordnance stores in quantity in 1757, and may even have remained there in sufficient quantities to be incorporated into Short Lands being set up in the 1768–1775 period.

One could speculate that the Ordnance "upgraded"

94

early Short Pattern muskets by swapping the old-fashioned rounded sideplate for the more current flat one, but an expensive upgrade of a nonfunctional piece of brass furniture makes no sense. It also would leave evidence on the inletting of wood stocks – evidence that is simply not there.

So, it must be acknowledged that the chasm between modern interpretations of what was written down in the 1770s and what was actually made and issued could turn out to be considerable. As with any ancient object, the careful study of the physical item versus its supporting documentation will often reveal a messy disconnect, and that appears to be the case here. Aside from all this scholarly debate, however, for the collector the answer is simple. Since all known examples of this musket have the same sideplates, we will simply call them all "Pattern 1769s" and leave it at that. The surviving muskets, if not all the surviving documentation, would seem to agree with us.

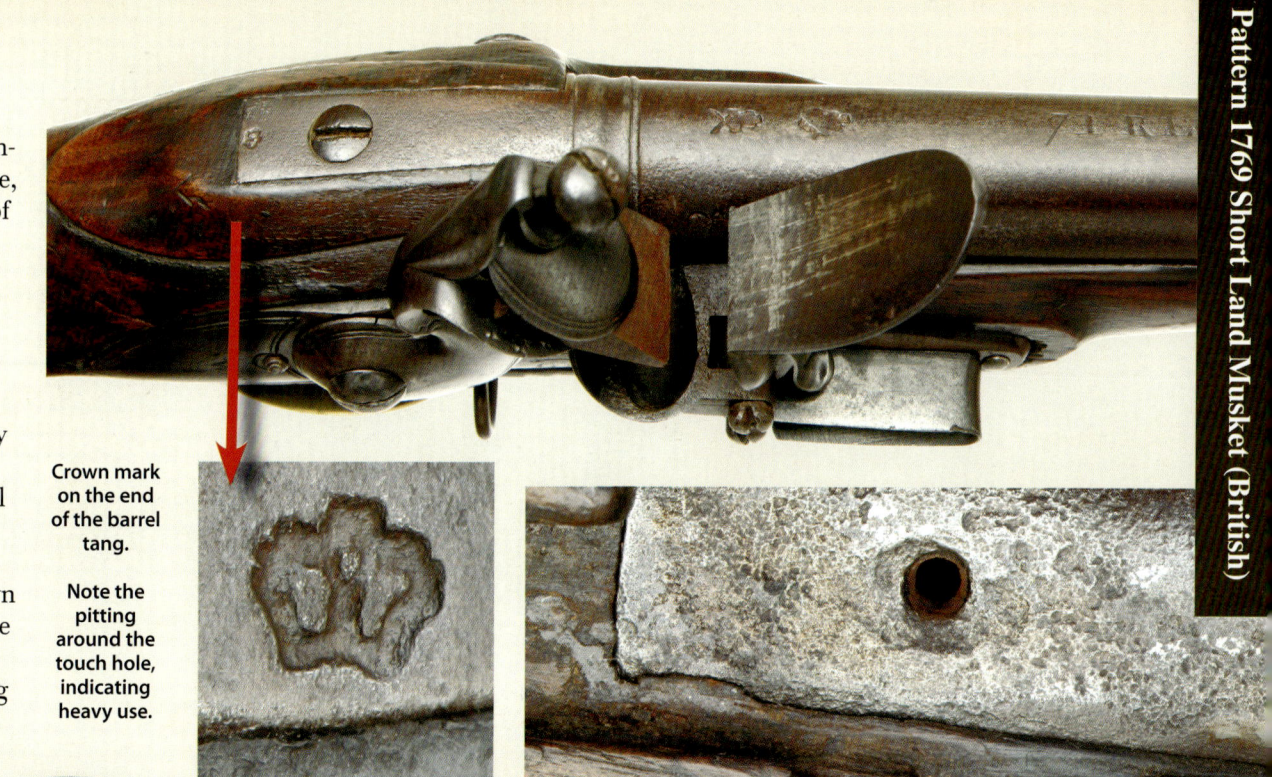

Crown mark on the end of the barrel tang.

Note the pitting around the touch hole, indicating heavy use.

Pattern 1769 Short Land Musket (British)

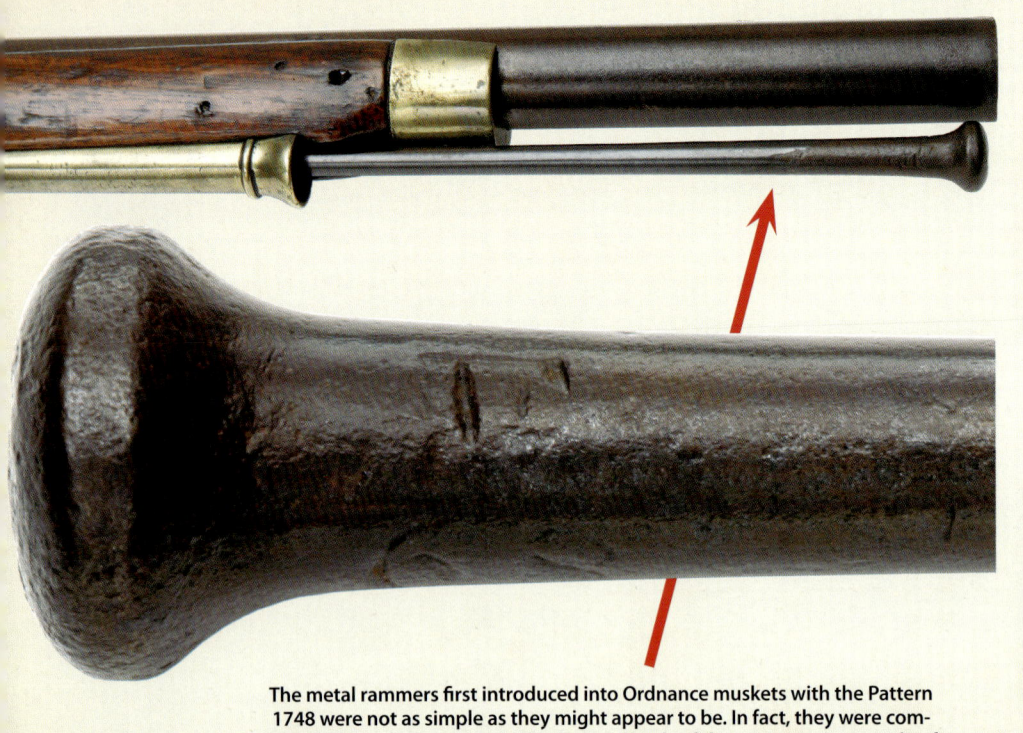

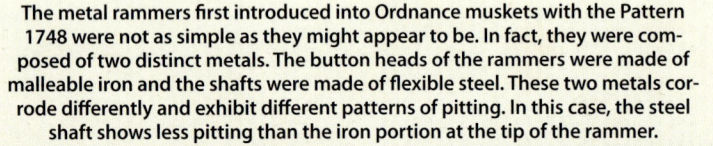

The metal rammers first introduced into Ordnance muskets with the Pattern 1748 were not as simple as they might appear to be. In fact, they were composed of two distinct metals. The button heads of the rammers were made of malleable iron and the shafts were made of flexible steel. These two metals corrode differently and exhibit different patterns of pitting. In this case, the steel shaft shows less pitting than the iron portion at the tip of the rammer.

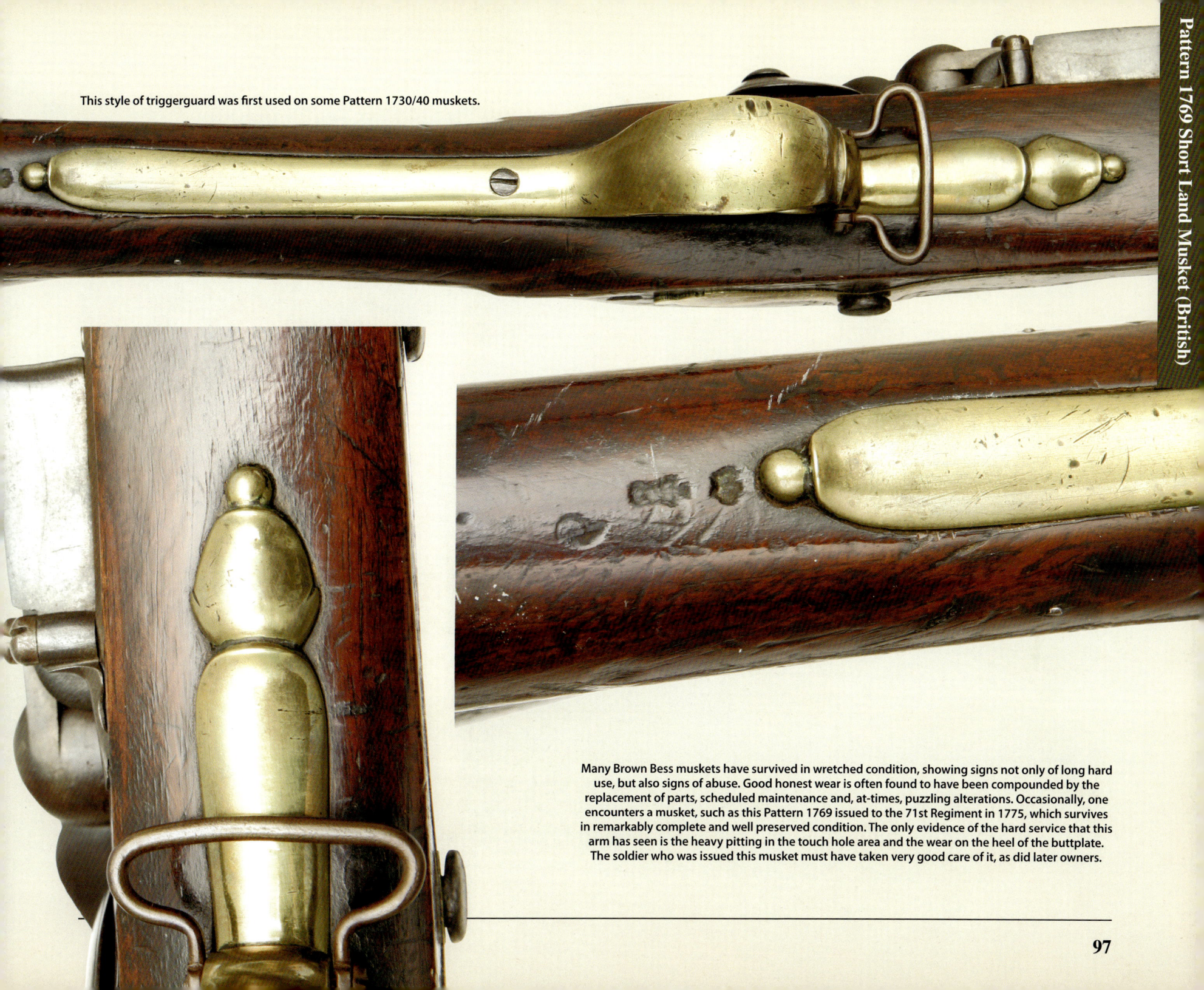

This style of triggerguard was first used on some Pattern 1730/40 muskets.

Pattern 1769 Short Land Musket (British)

Many Brown Bess muskets have survived in wretched condition, showing signs not only of long hard use, but also signs of abuse. Good honest wear is often found to have been compounded by the replacement of parts, scheduled maintenance and, at-times, puzzling alterations. Occasionally, one encounters a musket, such as this Pattern 1769 issued to the 71st Regiment in 1775, which survives in remarkably complete and well preserved condition. The only evidence of the hard service that this arm has seen is the heavy pitting in the touch hole area and the wear on the heel of the buttplate. The soldier who was issued this musket must have taken very good care of it, as did later owners.

97

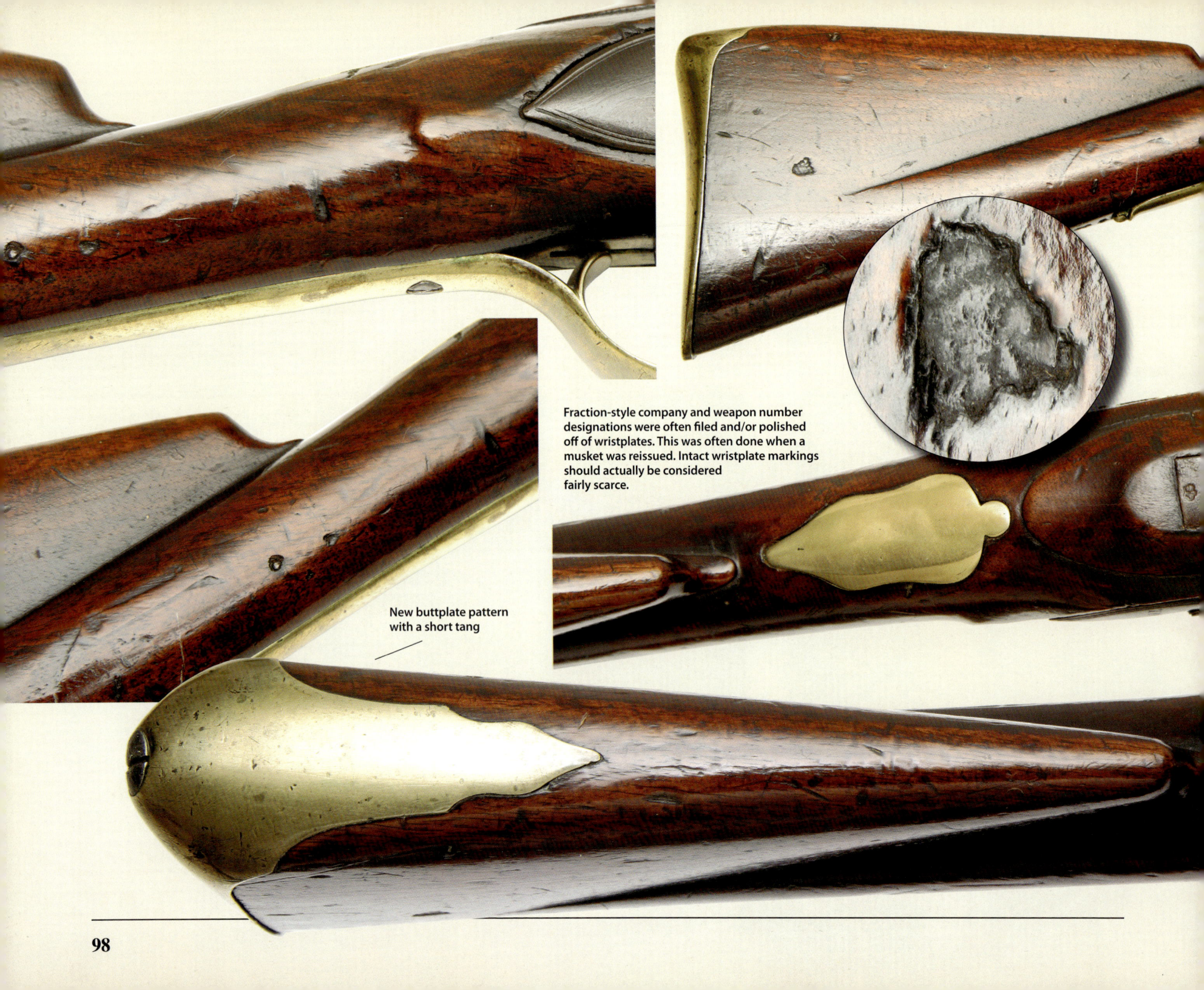

Fraction-style company and weapon number designations were often filed and/or polished off of wristplates. This was often done when a musket was reissued. Intact wristplate markings should actually be considered fairly scarce.

New buttplate pattern with a short tang

The flat sideplate used on these muskets was first introduced with the Pattern 1759 Militia or Marine musket. Maintaining the same outline of earlier Land Pattern sideplates, it required deeper inletting. Notice that when they drilled through the stock to put in the trigger pivot pin, they had the sideplate already installed and the drill went right through it (see red arrow).

Pattern 1769 Short Land Musket (British)

Assembly number on the screw

Simple carving around the barrel tang

Sideplate mortise.

Notice bulbous end to sideplate flat.

99

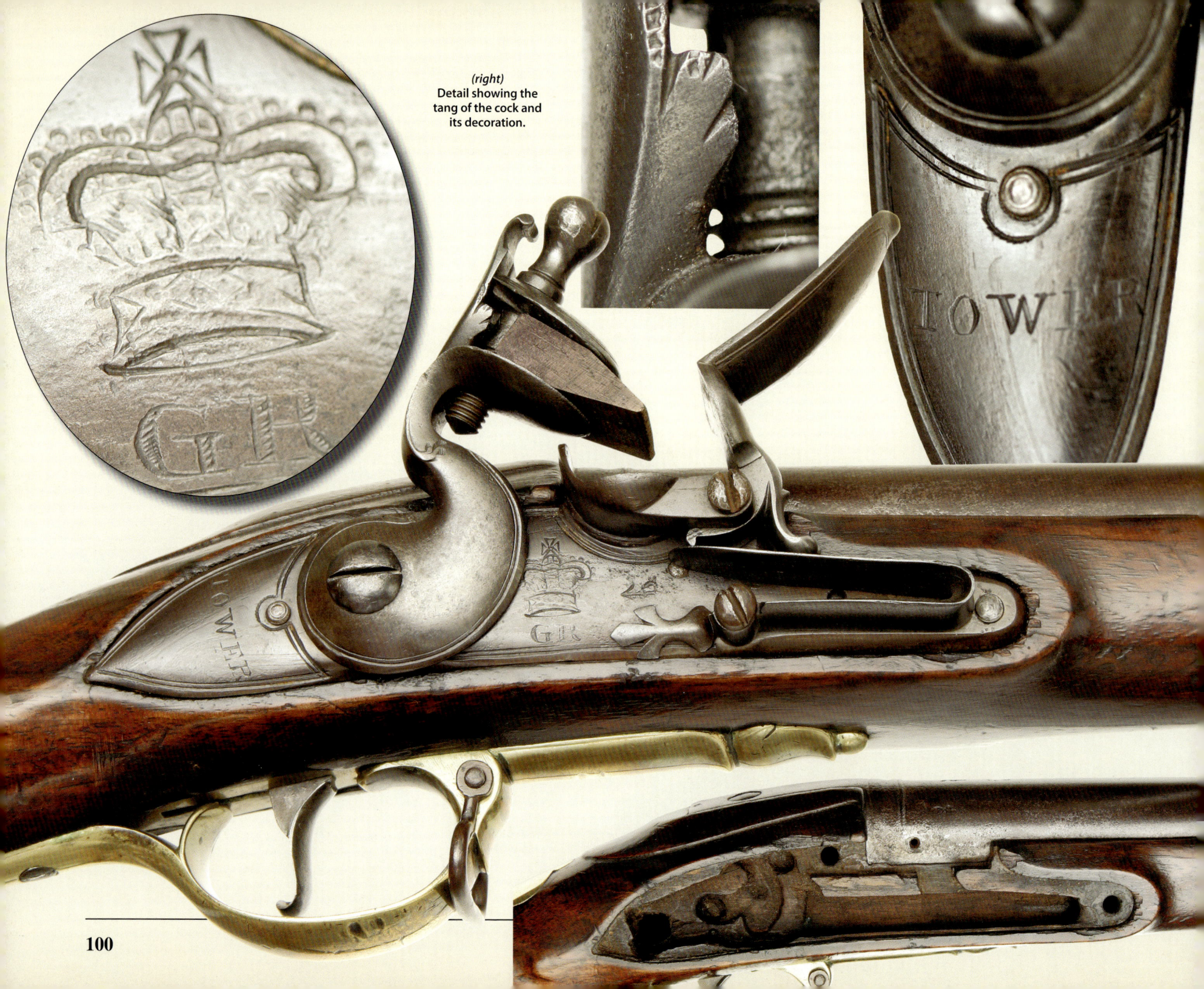

(right) Detail showing the tang of the cock and its decoration.

Pattern 1769 Short Land Musket (British)

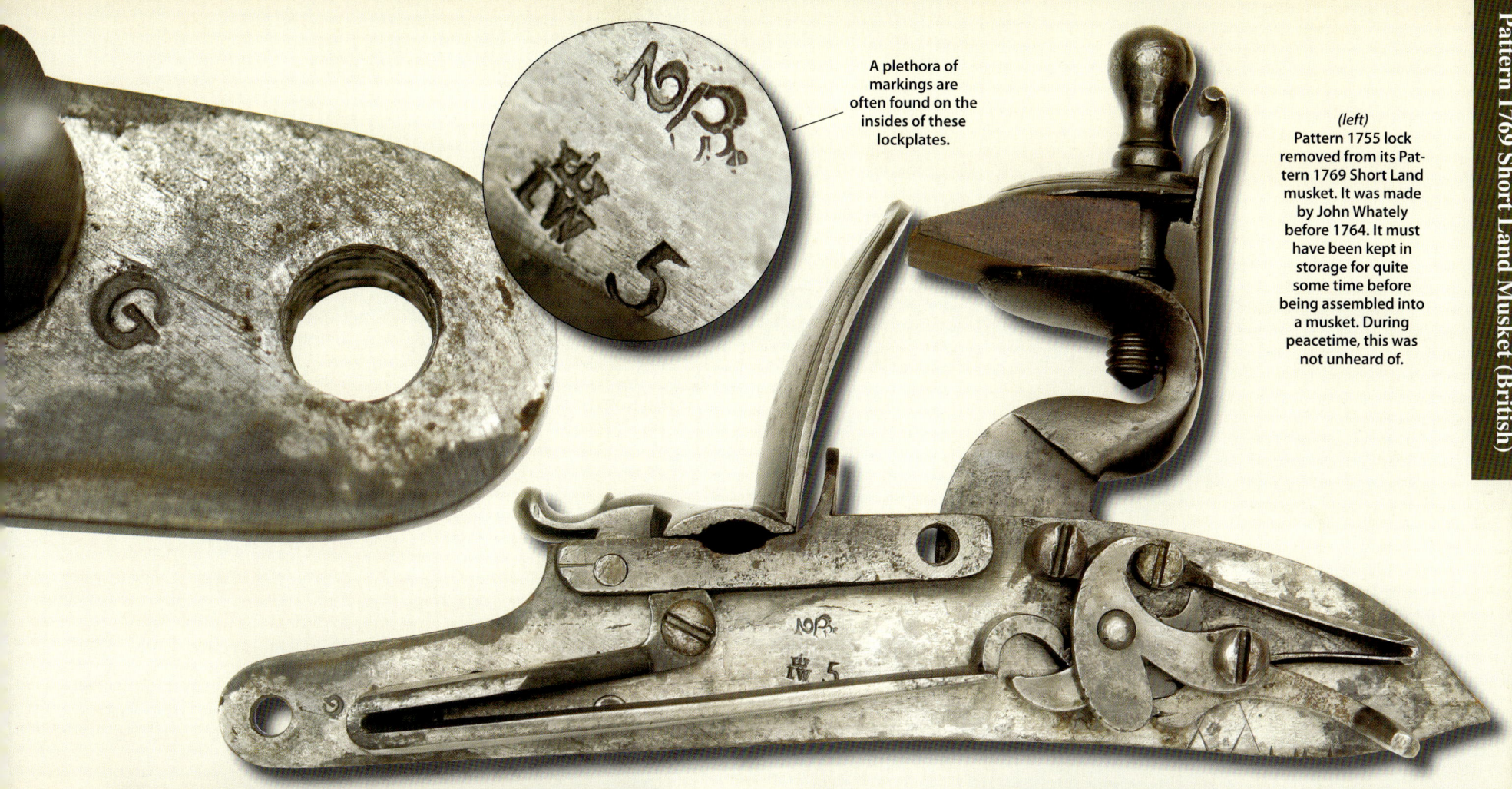

A plethora of markings are often found on the insides of these lockplates.

(left) Pattern 1755 lock removed from its Pattern 1769 Short Land musket. It was made by John Whately before 1764. It must have been kept in storage for quite some time before being assembled into a musket. During peacetime, this was not unheard of.

Lock details from two different Pattern 1755 locks on Pattern 1769 muskets. Notice the "WG" maker's mark of William Grice, the crowned "IW" lockmaker's mark, which indicates lock contractor John Whately. In both cases, the lock was inspected by Ordnance inspector number two and bears his crowned 2 mark.

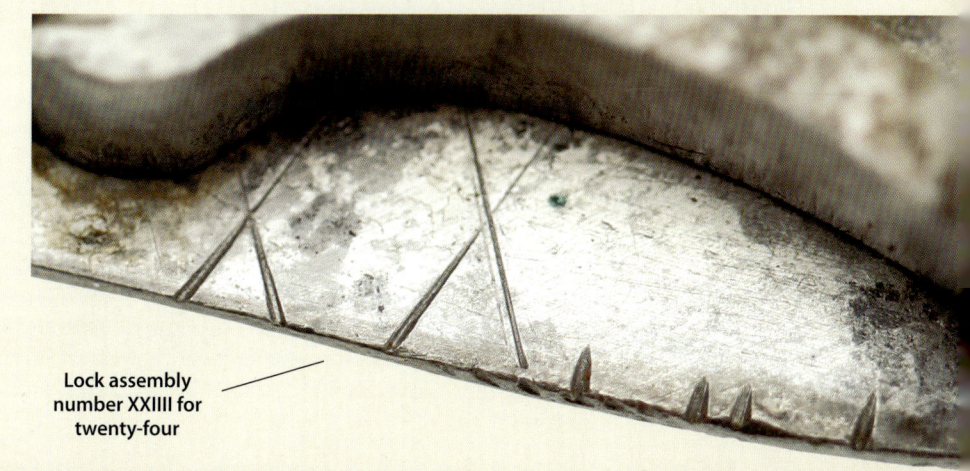

Lock assembly number XXIIII for twenty-four

101

THE PATTERN 1769 SHORT LAND MUSKET (IRISH)

An under-appreciated battle weapon of the Revolutionary War.

Rarity: Rare

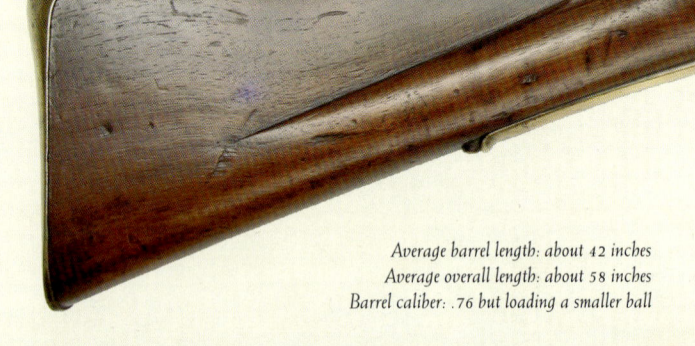

Average barrel length: about 42 inches
Average overall length: about 58 inches
Barrel caliber: .76 but loading a smaller ball

Since no one loves corny metaphors more than the authors of this book, we have previously compared Brown Besses to sexy girls, vintage muscle cars and heavy duty freight-hauling equipment. Not wanting to break with that noble tradition, we now offer the suggestion that the Irish Pattern 1769 is the "dark horse" of the Bess series, quietly avoiding the attention it surely warrants.

Production numbers for the Irish Pattern 1769 are dwarfed when compared to the British version – by a ratio of nearly five to one. This might lead the arms student to think the Irish musket was a scarcity during that period, and didn't play much of a part during the primary conflict of that era, the American Revolution. Nothing could be further from the truth!

Immediately preceding the outbreak of the war, there were almost as many British infantry regiments on the Irish Establishment as there were stationed in England and Scotland. Once hostilities began in the Spring of 1775, a huge number of these "Irish" units embarked for American service, and arms issues show that the Dublin Castle Pattern 1769 musket was the primary arm carried by those men. Authorized by warrants dating between 1773 and 1776, almost 8,000 of these muskets and bayonets were destined for North America. At least twenty regiments received them, including the 3rd, 9th, 15th, 19th, 20th, 22nd, 24th, 27th, 28th, 30th, 33rd, 34th, 40th, 44th, 46th, 49th, 53rd, 54th, 57th and 62nd Regiments. The sheer number of Irish-armed redcoats, combined with the combat records of their regiments, show that these Hibernian Pattern 1769s are of paramount importance when trying to understand "who carried what" during the Revolutionary War.

The Emerald Isle certainly couldn't match the industrial production of the British midlands and its highly specialized arms-making factories centered around Birmingham. The relatively small numbers of muskets and bayonets built for Dublin Castle couldn't always be assembled from locally procured parts. In order to fulfill their contracts with the Irish Ordnance, gun makers successfully petitioned for the right to acquire arms components from England, allowing for the importation of some 9,000 sets between late 1774 and late 1775. Included in these orders were Birmingham-made locks, barrels and bayonets – all supplied in a rough state.

To the collector or scholar familiar with these arms, the first trait of the Dublin Castle musket that pops into mind has to do with its overall quality when compared to one of its British sisters. It seems a little cliché to state that the latter is of a noticeably higher level of workmanship and finish, but that point really can't be debated, so it must stand as the paramount physical trait of this musket, regardless of its historical importance. With the components for thousands of Irish muskets being acquired in an unfinished state from sources in England, the lower quality of these Dublin Castle Pattern 1769s can't be completely blamed on Irish workmanship.

Structurally and mechanically, the Pattern 1755-style locks built into these guns are very close to the Tower-marked ones made for the British Ordnance, a phenomenon possibly ascribable to common origins. We've heard time and time again that the Irish locks carry "crude" engraving, which is easily distinguishable from that applied on British locks. More specifically, the engraving of "Dublin Castle" can be heavy handed with deeply cut triangu-

Production Period:
c.1770–1775

Total Production:
At least 14,772

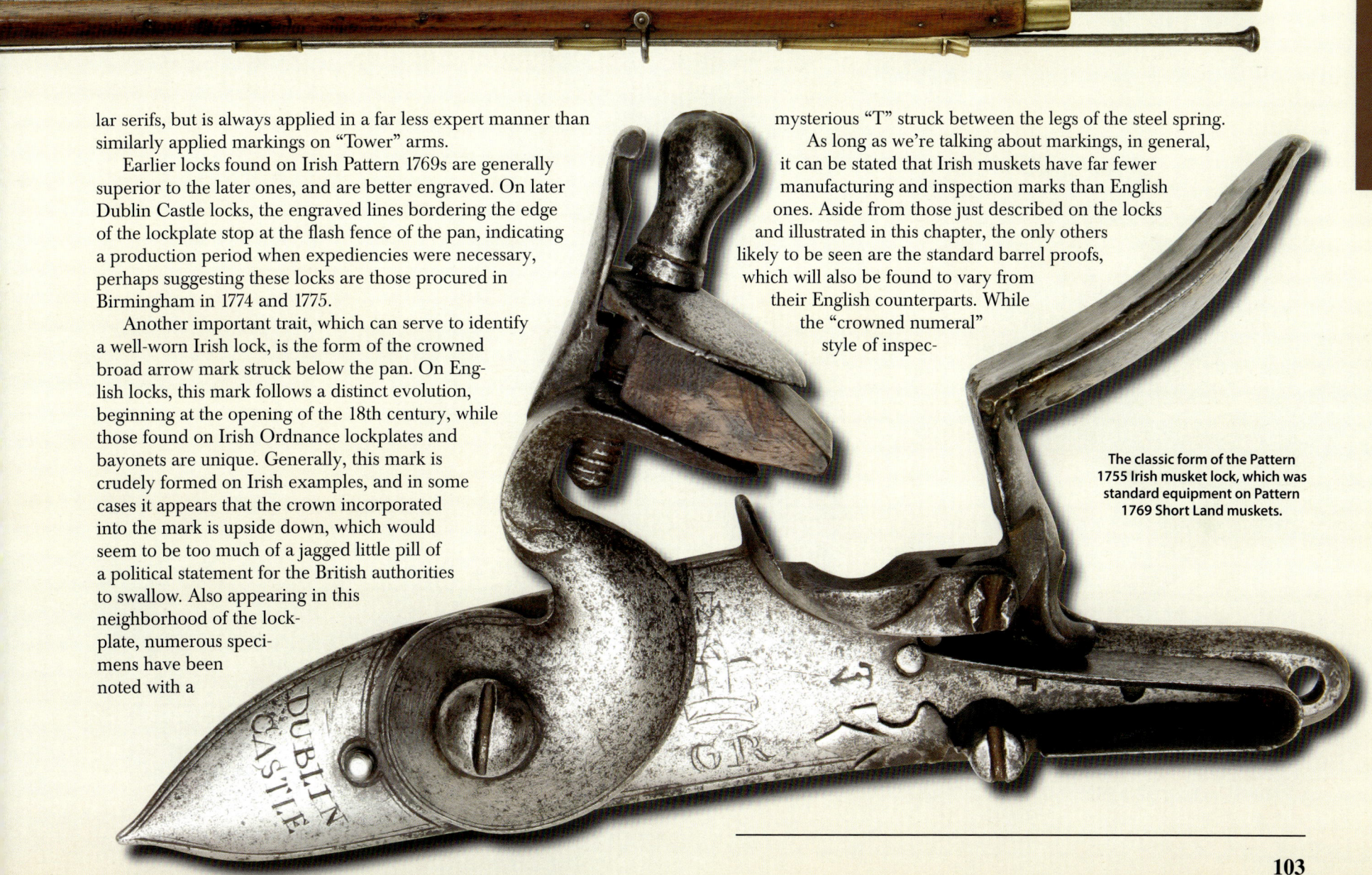

Pattern 1769 Short Land Musket (Irish)

lar serifs, but is always applied in a far less expert manner than similarly applied markings on "Tower" arms.

Earlier locks found on Irish Pattern 1769s are generally superior to the later ones, and are better engraved. On later Dublin Castle locks, the engraved lines bordering the edge of the lockplate stop at the flash fence of the pan, indicating a production period when expediencies were necessary, perhaps suggesting these locks are those procured in Birmingham in 1774 and 1775.

Another important trait, which can serve to identify a well-worn Irish lock, is the form of the crowned broad arrow mark struck below the pan. On English locks, this mark follows a distinct evolution, beginning at the opening of the 18th century, while those found on Irish Ordnance lockplates and bayonets are unique. Generally, this mark is crudely formed on Irish examples, and in some cases it appears that the crown incorporated into the mark is upside down, which would seem to be too much of a jagged little pill of a political statement for the British authorities to swallow. Also appearing in this neighborhood of the lockplate, numerous specimens have been noted with a

mysterious "T" struck between the legs of the steel spring.

As long as we're talking about markings, in general, it can be stated that Irish muskets have far fewer manufacturing and inspection marks than English ones. Aside from those just described on the locks and illustrated in this chapter, the only others likely to be seen are the standard barrel proofs, which will also be found to vary from their English counterparts. While the "crowned numeral" style of inspec-

The classic form of the Pattern 1755 Irish musket lock, which was standard equipment on Pattern 1769 Short Land muskets.

103

The engraving on these Irish locks varies from crude to cruder when compared to their British counterparts.

tion mark is seen from time to time (mostly on the inside of lockplates), marks in the stock, rammer channel and the flip sides of the brass mounts are dispensed with. Also deemed unnecessary were the Ordnance Storekeeper's marks struck into the lock-side buttstock on English arms.

After handling enough English and Irish Besses to get a general feel for them, a number of other differences also become apparent. Irish stocks often have a pale, lighter color of walnut (sometimes with an orangey hue), while English ones tend to have a deeper, more reddish color. Put one of each side to side, and the Dublin Castle musket will often look slightly anemic. Tower muskets have forestocks that taper nicely as their sides arc gracefully upward to meet the barrel channel, while Irish muskets often have a step there, giving the forestock a clunky, awkward appearance (see the close-up photo on page 107, center right). And while the beefiest area of any musket stock should be the breech-lock-sideplate section, on Irish arms this section is often surprisingly slight.

Contrasting the established "Tower of London" style of regimental, company and weapon number markings, it should be no surprise that those engravings differ significantly when encountered on Irish guns. Certainly not a testament to the skill level of their engravers, these widely diverse unit markings are indicative of the many ways they could be – and were – applied at the regimental level after they had been issued. Perhaps erroneously referred to as "in the field" markings, these engravings were

Pattern 1769 Short Land Musket (Irish)

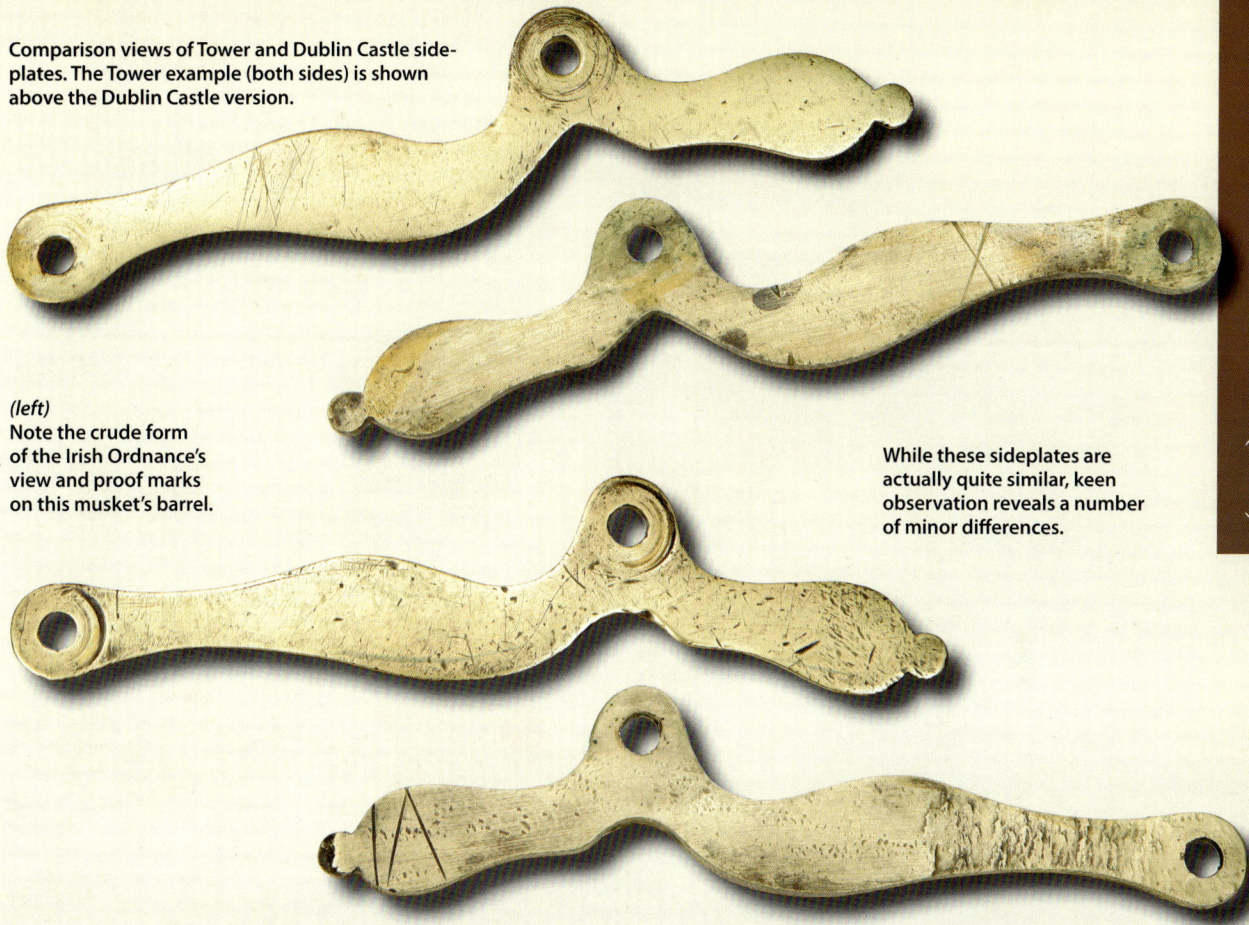

Comparison views of Tower and Dublin Castle sideplates. The Tower example (both sides) is shown above the Dublin Castle version.

(left) Note the crude form of the Irish Ordnance's view and proof marks on this musket's barrel.

While these sideplates are actually quite similar, keen observation reveals a number of minor differences.

applied at the expense and whim of the regiment responsible for them. Some were marked by skilled regimental armorers and some were executed by locally employed engravers. Beauty and uniformity apparently didn't matter, just as long as the markings themselves were readable enough to clearly establish ownership. Thus, markings will be found along the top of the barrel in imitative "Tower" fashion, on the wristplate and even on the bow of the triggerguard. It can be assumed that the wristplate would be the most popular location for these marks, because it was made of soft, easily engraved brass and could quickly be removed from the musket by undoing just one screw. This is borne out by surviving specimens.

To date, of the twenty British regiments that fought the American Revolution carrying Dublin Castle Pattern 1769s, examples marked to ten of them are known. Other muskets, like one shown in this chapter, have had their regimental markings filed off their wristplates by later owners. Marked or not, the unsung Irish Pattern 1769 Short Land musket is one of the most consequential arms of the War of Independence, and deserves much more attention from scholars and collectors than it has received so far. Perhaps one day they will shed their dark horse status and be raised aloft...placed so deservedly in the Pantheon of history's great weapons – even if they are a little bit ugly.

105

Note the sloppy brazed refacing of the frizzen.

Exceptionally worn screw slot

The new face of this repaired frizzen is partially delaminated — essentially, it peeled off.

Pattern 1769 Short Land Musket (Irish)

This area of the stock is much thinner on some Irish muskets than on the British versions.

The lip of the barrel bed has a distinctive, flat form on many Irish muskets, while British muskets are tapered in this area.

A variety of views showing the Irish Pattern 1769's sideplate and its mortise in the stock.

Note that Irish stocks frequently use a type of walnut with a lighter appearance than what is normally expected on British Ordnance stocks.

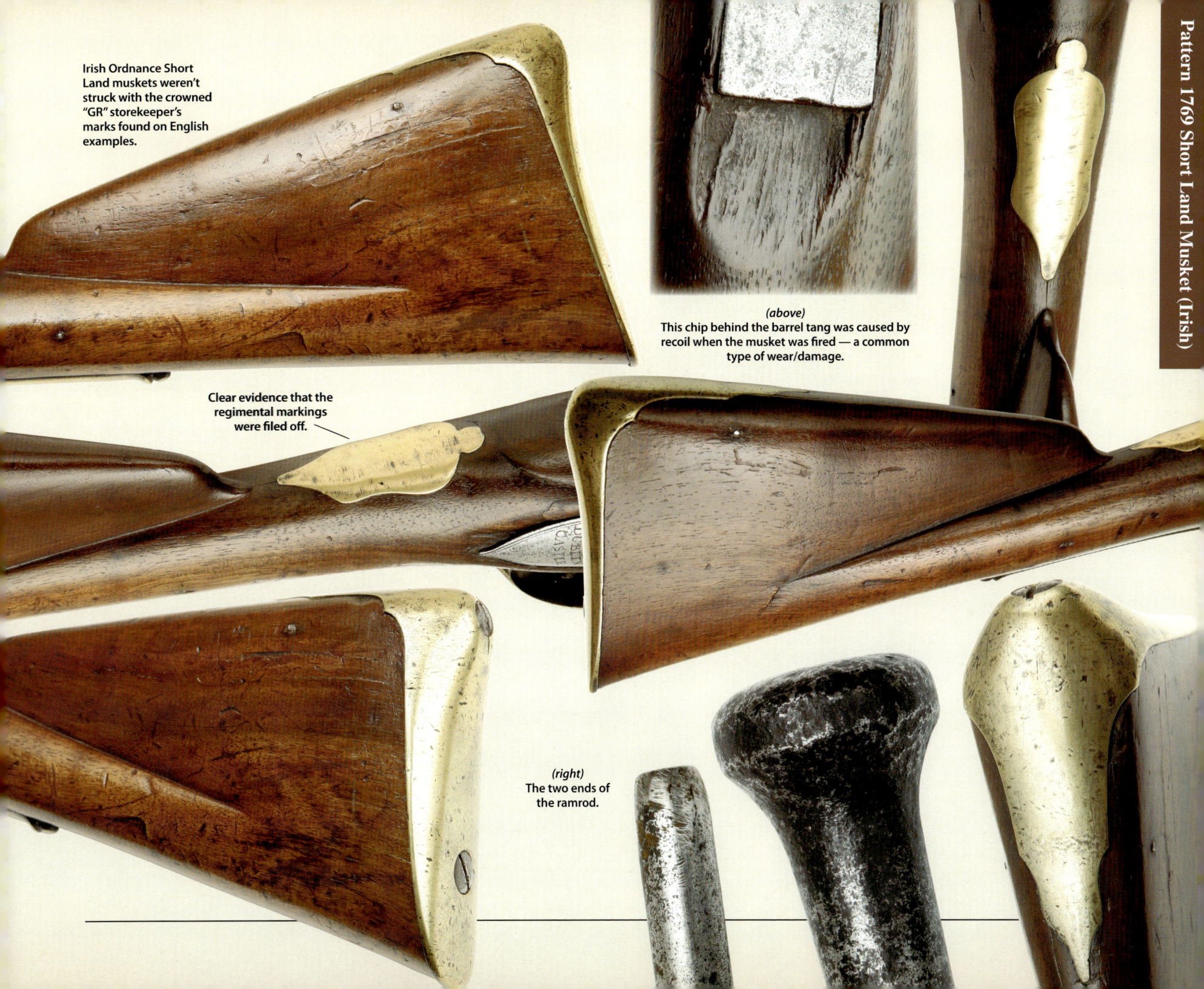

Irish Ordnance Short Land muskets weren't struck with the crowned "GR" storekeeper's marks found on English examples.

(above)
This chip behind the barrel tang was caused by recoil when the musket was fired — a common type of wear/damage.

Clear evidence that the regimental markings were filed off.

(right)
The two ends of the ramrod.

Pattern 1769 Short Land Musket (Irish)

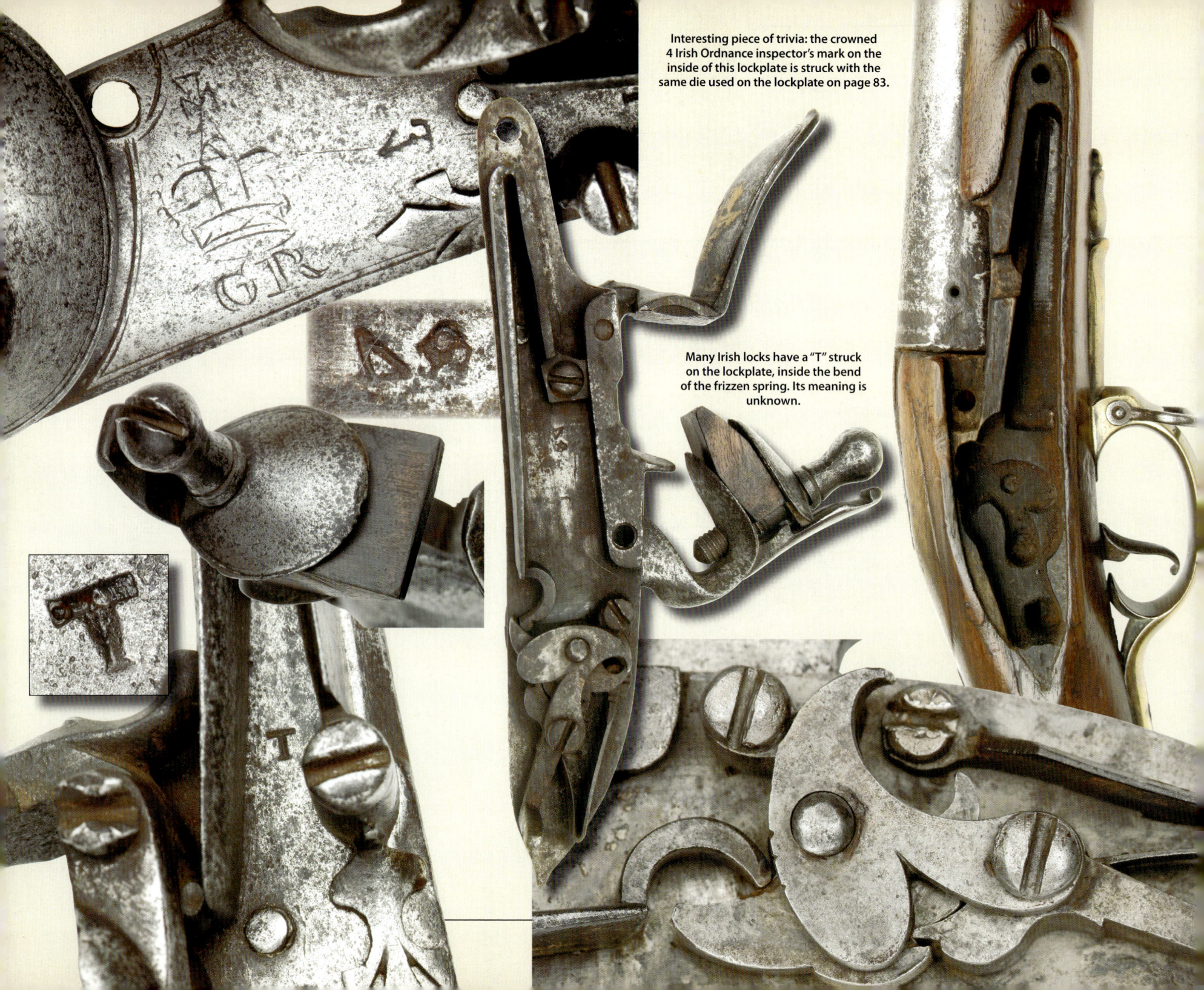

Interesting piece of trivia: the crowned 4 Irish Ordnance inspector's mark on the inside of this lockplate is struck with the same die used on the lockplate on page 83.

Many Irish locks have a "T" struck on the lockplate, inside the bend of the frizzen spring. Its meaning is unknown.

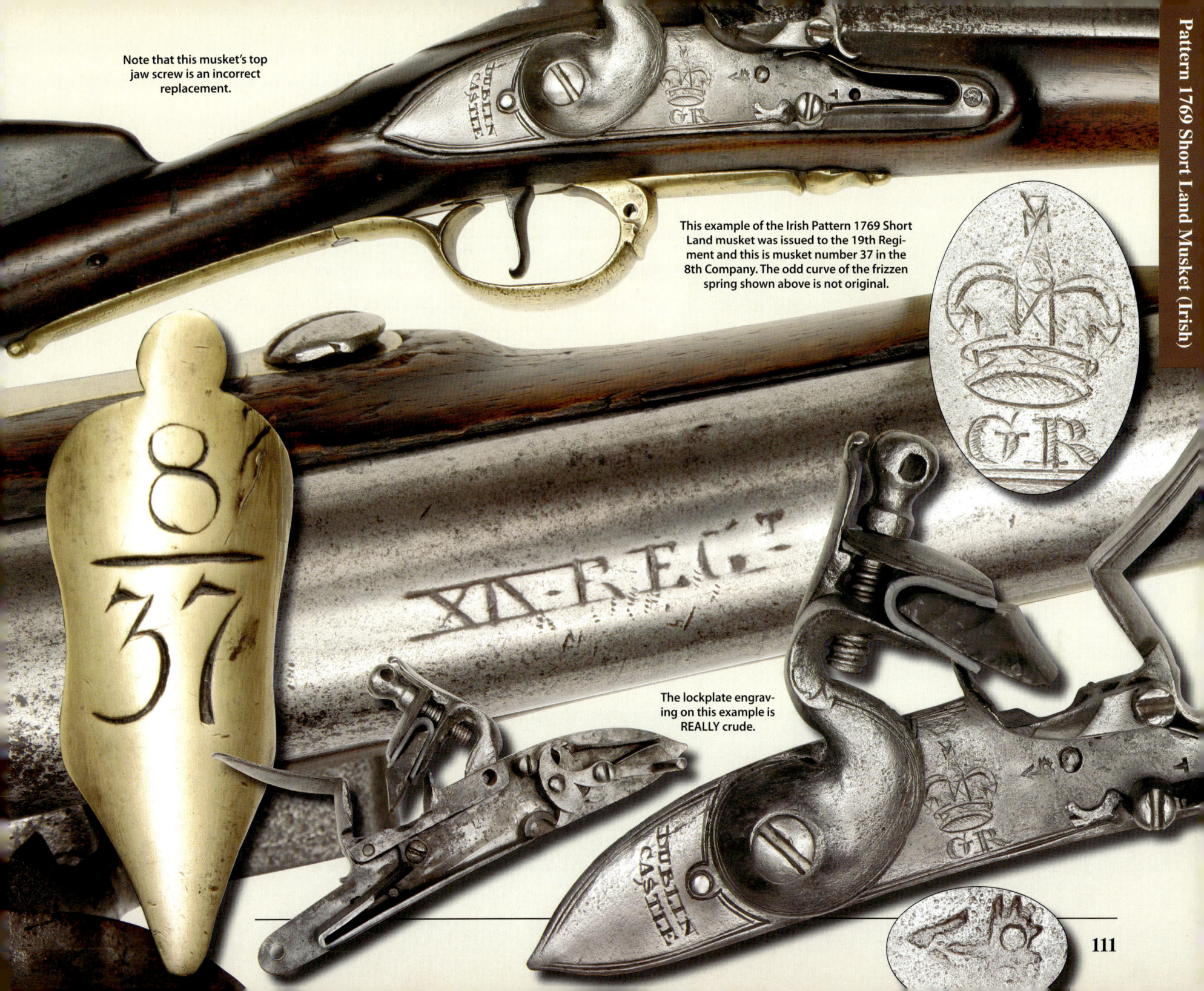

Pattern 1769 Short Land Musket (Irish)

Note that this musket's top jaw screw is an incorrect replacement.

This example of the Irish Pattern 1769 Short Land musket was issued to the 19th Regiment and this is musket number 37 in the 8th Company. The odd curve of the frizzen spring shown above is not original.

The lockplate engraving on this example is REALLY crude.

111

THE PATTERN 1777 SHORT LAND MUSKET

A new lock design sets this long-lived Short Land Pattern apart.

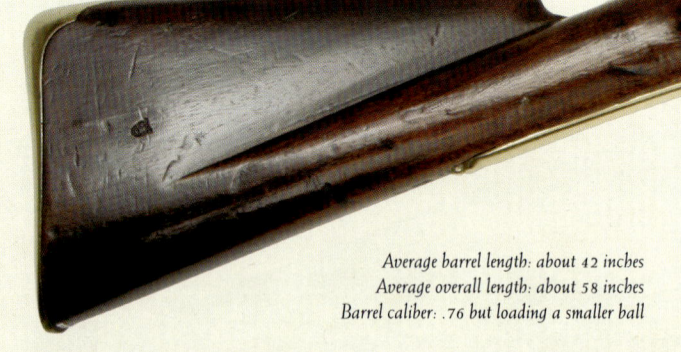

Rarity: Scarce

Average barrel length: about 42 inches
Average overall length: about 58 inches
Barrel caliber: .76 but loading a smaller ball

By the end of the American Revolution, almost half of the 350,000-plus Short Land muskets procured by both the English and Irish establishments were variants of a pattern introduced in early 1777. Economic factors coupled with a slight improvement in functionality gave birth to this long-standing and sturdy service musket.

The new musket's distinct traits were clearly spelled out in 1970 by Anthony Darling in his booklet *Red Coat and Brown Bess*, and they are listed in the blue table to the right. As can be seen, the most dramatic change is a new lock, dubbed the "Pattern 1777" by arms researcher De Witt Bailey. Its new mechanism was a significant break from previous Ordnance lock patterns.

The last item in the blue table has actually become a bone of contention amongst arms students. Bailey has been able to document just 15,000 of these wide-mouthed second rammer pipes coming into Tower stores; they arrived between late January and late March of 1779. This would seem to suggest that only 15,000 Ordnance System muskets could have been made using this particular style of pipe, making it an extremely rare feature. Indeed, according to Bailey's musket pattern date system, Short Land muskets with the Pattern 1777 lock and a straight-sided second pipe are termed "Pattern 1777" muskets, while those made with the wider mouthed pipe are called the "Pattern 1779" musket. There's just one problem. *Every single original-condition example that we have ever seen* has had this supposedly rare second pipe. Based upon bayonet production, we can safely assume that literally tens of thousands of muskets were manufactured for the Ordnance between 1777 and 1779, and not a single one has been found having the supposedly much-more-common straight second pipe. What are we to make of this conundrum?

Let's examine some of the other evidence. The Ordnance minutes of 2 May 1777 recorded not only the existence of this wider-mouthed pipe, a creation of John Pratt, but also the Board's favorable opinion of the design and its practical application. Since this new pipe

Features of the Pattern 1777 Short Land Musket

- Two screw heads are visible behind the cock on the tail of the lockplate. This is caused by a shortened sear spring.

- The tang of the cock is shaped like a solid pillar topped by a rudimentary forward curl.

- The top jaw and the back of the steel are devoid of double-line engraving.

- The top jaw is notched to fit around the sides of the tang of the cock.

- The top jaw screw is simpler than on earlier locks and is pierced with a prominent, large hole for tightening.

- The steel spring finial is now shaped like an elongated pear.

- The second rammer pipe (below the long trumpet pipe) has acquired a taper with a wider mouth.

Production Periods:
First Period: 1777–1782
Second Period: 1792–1804

Total Production:
Less than 260,835 total (with a very small number made for Dublin Castle). From 1777–1782, Ordnance made fewer than 157,639. Contractors also supplied Pattern 1777 variants complete: 14,508 from 1778–1780 and 88,688 from 1792–1804.

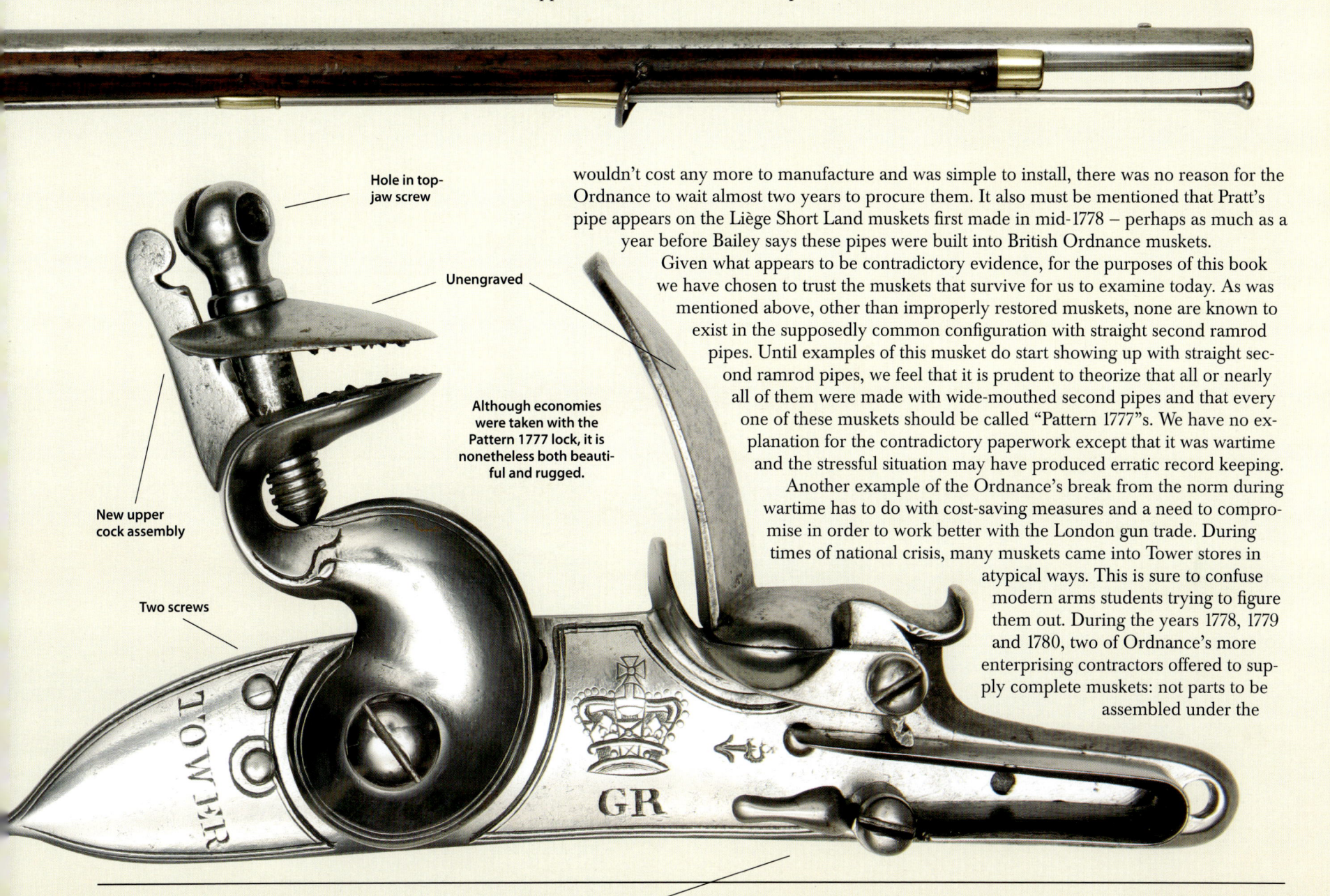

Pattern 1777 Short Land Musket

Hole in top-jaw screw

Unengraved

Although economies were taken with the Pattern 1777 lock, it is nonetheless both beautiful and rugged.

New upper cock assembly

Two screws

Simplified spring finial

wouldn't cost any more to manufacture and was simple to install, there was no reason for the Ordnance to wait almost two years to procure them. It also must be mentioned that Pratt's pipe appears on the Liège Short Land muskets first made in mid-1778 – perhaps as much as a year before Bailey says these pipes were built into British Ordnance muskets.

Given what appears to be contradictory evidence, for the purposes of this book we have chosen to trust the muskets that survive for us to examine today. As was mentioned above, other than improperly restored muskets, none are known to exist in the supposedly common configuration with straight second ramrod pipes. Until examples of this musket do start showing up with straight second ramrod pipes, we feel that it is prudent to theorize that all or nearly all of them were made with wide-mouthed second pipes and that every one of these muskets should be called "Pattern 1777"s. We have no explanation for the contradictory paperwork except that it was wartime and the stressful situation may have produced erratic record keeping.

Another example of the Ordnance's break from the norm during wartime has to do with cost-saving measures and a need to compromise in order to work better with the London gun trade. During times of national crisis, many muskets came into Tower stores in atypical ways. This is sure to confuse modern arms students trying to figure them out. During the years 1778, 1779 and 1780, two of Ordnance's more enterprising contractors offered to supply complete muskets: not parts to be assembled under the

113

usual Ordnance System, but complete, finished weapons. The Ordnance Department was not in a position to be picky or stubborn, and almost 20,000 arms were provided in this way by contractors James Hirst and John Pratt. The vast majority (almost 75%) of these weapons are believed to be close copies of the Pattern 1777 musket. Of these 14,500+ Pattern 1777 copies, none have yet been identified positively. All that can be said about these weapons is that they probably would have carried fewer inspector's marks than muskets assembled under the Ordnance System.

Overlapping with the production of the India Pattern musket for at least 11 years, the Pattern 1777 takes the cake for having the longest production run in the Land Pattern musket series. Since so many of them were made during the post-1792 period (about 30% of the Pattern 1777 total), one must be diligent when determining which production run a particular specimen belongs to. A sure tell-tale sign is the manner in which the "Crown GR" and "TOWER" were applied to the lockplate. If they are engraved, then it should be "first period" (1777–1782), and if they are stamped, then it's "second period" (1792–1804) for sure. Other clues can be had by researching the dates of any contractors' names or initials found stamped into the various components of the musket. Such is the case with the musket illustrated here. With the name "WALLER" clearly stamped into the rammer channel, we can date this musket to before 1781, when James Waller ceased "setting

Even though the Pattern 1777 musket dispensed with certain features in the name of savings, great care was still taken in the inletting of the lock as can be seen here. The crowned 4 inspector's mark struck on the barrel probably represents the Ordnance's approval of the touch hole.

As typically seen, this Ordnance storekeeper's mark appears muddled. The stampings are never very precise or easy to read due to the nature of the wood into which they were struck.

The name "WALLER" is struck into the ramrod channel of this musket. James Waller only set up Pattern 1777 muskets for the Ordnance until 1781, which nicely dates the manufacture of this example to the height of the Revolutionary War.

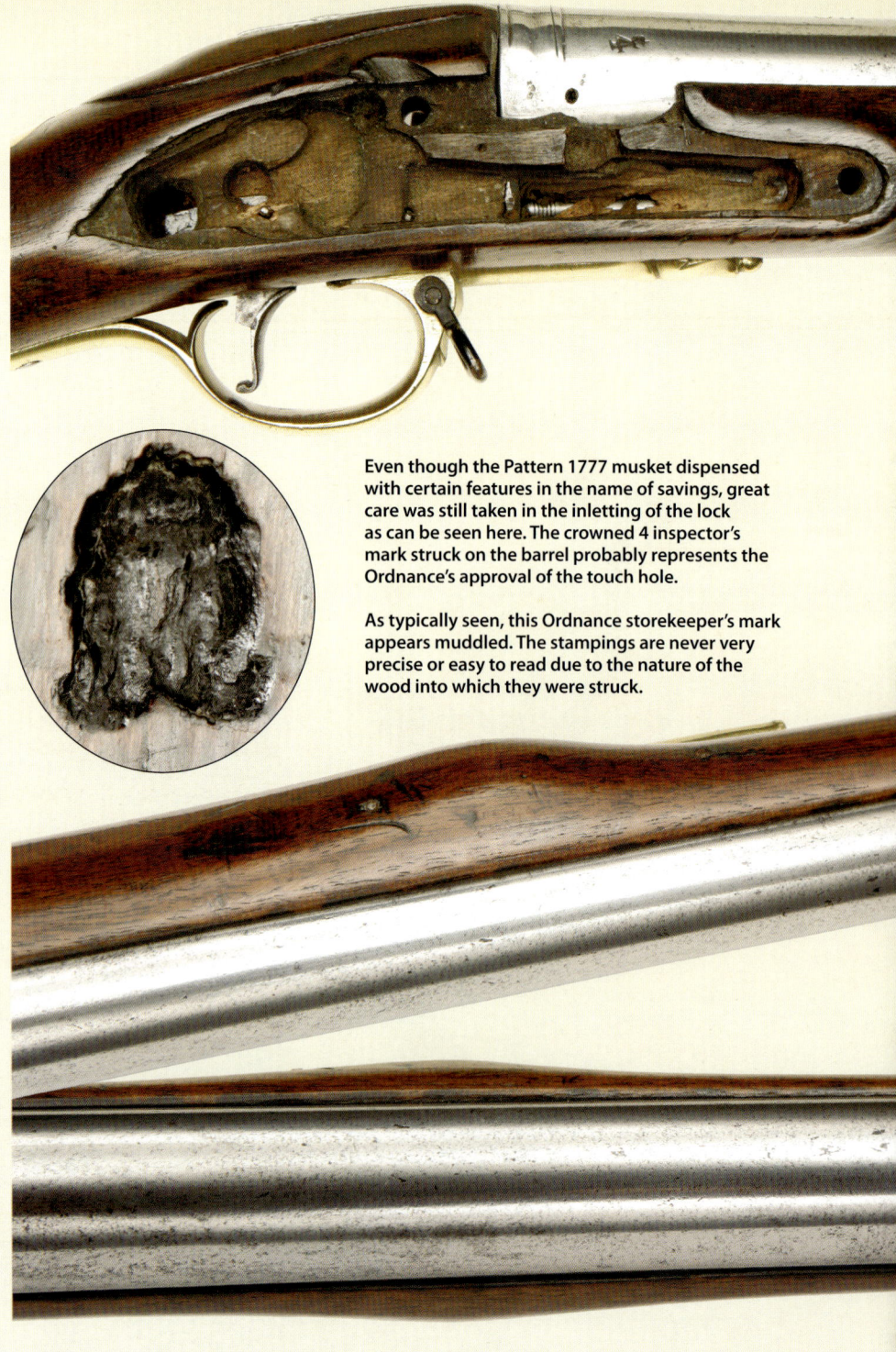

up" arms for the Ordnance. Because of this little fact, we can be sure that we are looking at a splendid example of a wartime-produced Pattern 1777.

The use of the Pattern 1777 musket during the American Revolution is yet another good subject for lively debate. With no substantial clues (either in the written or archaeological records), we must look to the guns themselves for hints of wartime service. Pattern 1777s have been seen with regimental markings for the 14th, 46th, 54th, 60th and 79th Regiments – to name but a few. Additionally, a number of Pattern 1777s are marked for issue to the Queen's Rangers (a.k.a. the 1st American Regiment), a famed Loyalist unit that was disbanded after the war. The problem is that every one of these British units existed after the Revolution and might have received their Pattern 1777 muskets then. And the Queen's Rangers were reincarnated for service from 1791 until 1802, creating a similar opportunity for confusion.

So, as it stands, these sexy regimental markings actually offer no real proof of Revolutionary War service. In the case of the Queen's Rangers, it seems hard to believe that an American regiment would receive top-of-the-line muskets from the Ordnance during the Revolution – earlier than the bulk of the British Army then serving in the colonies. Furthermore, since these regimental markings are applied in exact Tower fashion, it perhaps seems more likely that these Pattern 1777s were supplied during peacetime when fully tricked-out arms could be provided for far-flung Canadian garrison units.

However, it must be remembered that the two daughters of the Pattern 1777, the Liège Bess and the Pattern 1779-S, definitely served in North America during the War of Independence. So why not the mother, herself? Therefore, the Pattern 1777 must be counted as a British weapon of the War of Independence until proven otherwise. However, attributing Revolutionary War service to any particular example remains problematical.

The wide-mouthed second ramrod pipe that first appeared on the Pattern 1777 musket.

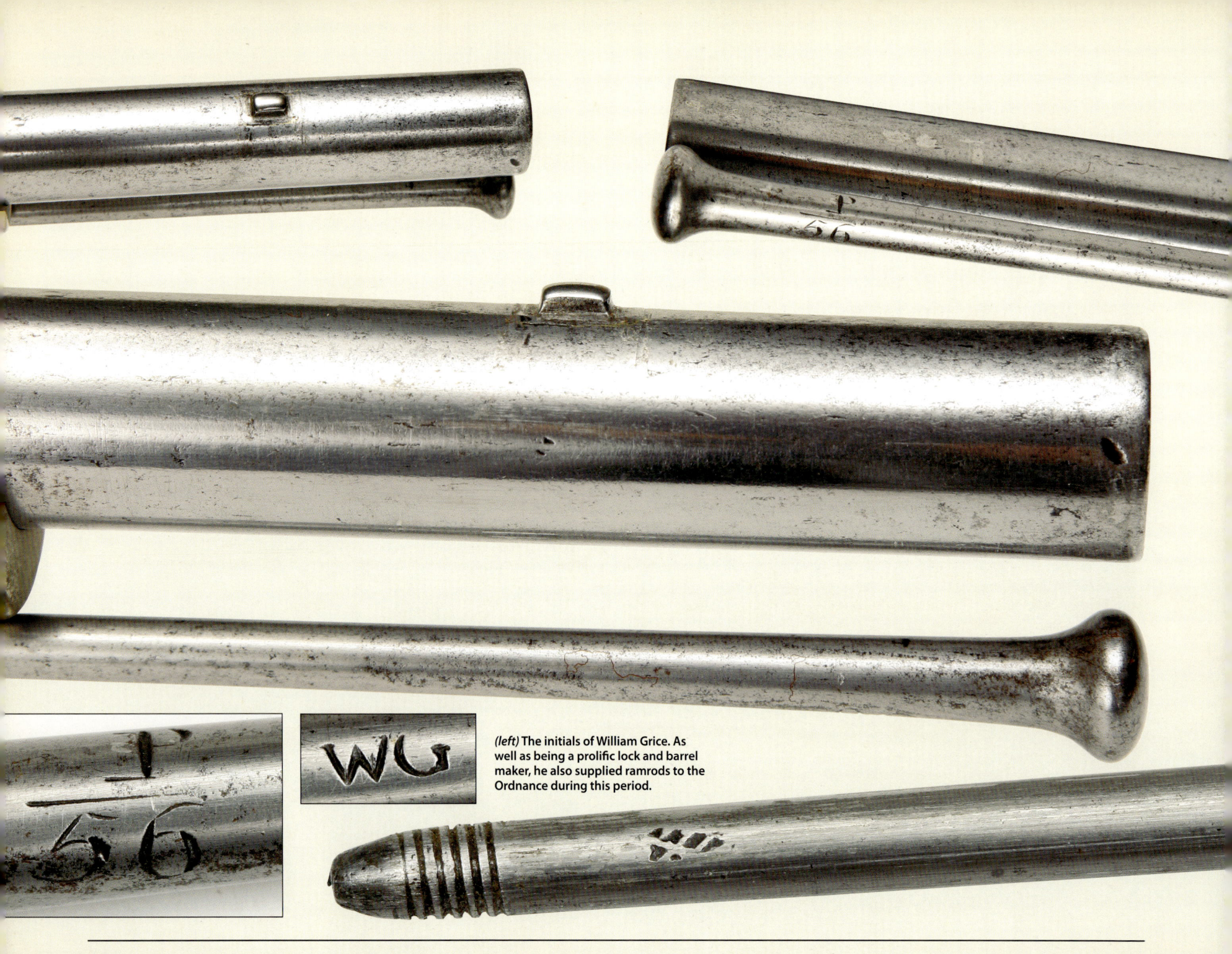

(left) The initials of William Grice. As well as being a prolific lock and barrel maker, he also supplied ramrods to the Ordnance during this period.

116

Pattern 1777 Short Land Musket

117

118

Pattern 1777 Short Land Musket

This crown mark appears on the inside of the triggerguard bow.

Unlike all of the previous patterns of muskets covered in this book, the Pattern 1777 Short Land musket can be located today in top condition. Dozens of these muskets can be found in museum collections on both sides of the Atlantic. During the period following World War II, quite a few examples of the Pattern 1777 came onto the market when a number of large displays of arms were broken up and sold from stately country homes in the countryside of Great Britain. It is believed that the bulk of these muskets were formerly the armaments of local militia units rather than ever having seen service in any of the regular army regiments — which would help to explain why so many of them have come down through the years in near-new condition.

119

While the large crown over "GR" on the lockplate is engraved, the pearls along the top of the crown are punched.

"Tower", crown over "GR" and crowned broad arrow (pointing sideways) — all common lockplate markings on these muskets.

Both lock screws and the inside of the sideplate are marked with the musket's assembly number of VIII.

120

Pattern 1777 Short Land Musket

Lock contractor Benjamin Willetts, whose last contract was in 1780

Pierced top-jaw screw

The descent arrest, a feature that stopped the cock from moving too far forward. Without this, the mainspring would be released, breaking the musket's stock.

The shorter sear spring on the Pattern 1777 lock is the reason why we see two screws protruding through the lockplate behind the cock on these locks (see page 101 for comparison).

(left) The unengraved frizzen on Pattern 1777 locks.

121

LIÈGE SHORT LAND PATTERN MUSKETS

Ordnance orders cheap muskets from the Low Countries.

Rarity: Scarce

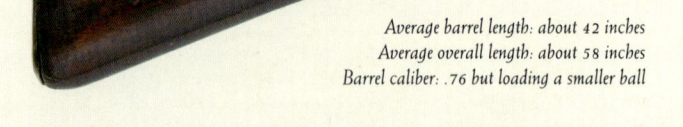

Average barrel length: about 42 inches
Average overall length: about 58 inches
Barrel caliber: .76 but loading a smaller ball

The 200,000+ Short Land muskets that poured into British Ordnance stores in England and Ireland during the war years of 1775 through 1782 were simply not sufficient to meet the demand. So, as they had done since the age of the Tudors, British authorities turned once again to commercial arms makers on the European Continent for an additional supply of muskets and bayonets.

However, instead of simply buying whatever was available for sale in the Low Countries of Belgium and the Netherlands (which had been the normal approach), this time the Board of Ordnance sent over specific patterns for the arms to be manufactured. Working through a succession of middlemen, a current Short Land Pattern musket was sent over for duplication, which was delivered on February 20, 1778. This pattern musket would be used as the model for all these Liège muskets. While there is some debate as to which musket was sent, it is very likely that it was of the Pattern 1777 type.

Contracts were arranged through middlemen, with many of the actual muskets coming from Liège gunmaker and merchant Jean Claude Niquet. A review of the known documentation is presented in De Witt Bailey's article "Liège Muskets in the British Army, 1740–1780" in the book *Aspects of Dutch Gunmaking*. The bottom line, however, is that muskets came from a tangled mess of Low Countries sources (collectively referred to as "Dutch" by the British) with a minimum of 76,000 arms being obtained from Liège between 1778 and 1783. According to Bailey, a maximum total of 110,000 may have been delivered.

As a group, the resulting muskets do vary. Given the number of deliveries and the "cottage industry" nature of the Liège arms trade, this should not surprise us. And while a couple of traits of the Pattern 1769 musket's lock (the Pattern 1755) have been noted on most examples, the simultaneous appearance of several Pattern 1777 features leads one to conclude that the latter pattern had to have been their prototype. Many of the Liège locks are similar to the English Pattern 1755 lock only in that one screw protrudes through the lockplate behind the cock, and there is double-line engraving on the back of the steel (when it survives). But that is where the similarity stops. The way that the comb of the cock is notched into the top jaw and the "plain lobe" steel spring finial are direct copies of those features on the Pattern 1777 lock.

- The double line engraved around the lockplate and the bottom of the cock are sometimes, but not always, set further back from the edges they follow, and don't continue under the pan.

- Similarly, the crown over the "GR" on the lockplate will vary from one example to another, having either a single or double-arched top. The crowns on British and Irish examples always have double arches on the top.

- The word "TOWER" is engraved with broader letters cut with crosshatched uprights. This is most noticeable on the "W".

- No "broad arrow" appears with the "crown" ownership stamp below the pan. Furthermore, this mark is usually oriented 90 degrees counterclockwise from the English norm (with the broad arrow pointing towards 9 o'clock).

Production Period:
1778–1783

Total Production:
110,000 maximum

Less commonly seen, some examples have been noted with two screws protruding from the tail of the lockplate, breaking any idiosyncratic structural ties with the 1755 Pattern lock. Clearly, the new Ordnance lock instituted the previous year is what went to Liège.

As with the engraving on Irish Ordnance locks, the engraving applied on these Liège-made mechanisms has been described as crude or heavy-handed when compared to the superior products of the British Ordnance system. And it is true. There are a number of specific things to look for on these "crude" markings that can be used to help identify a Liège product, and these are covered in the blue table on the previous page. It should also be noted that while the large crown engraved on the lockplate's face varies in its minor details from example to example, there appear to be two basic styles – one being more heavy handed than the other. Both types will be illustrated further on in this chapter for comparison. These different styles of lockplate engraving do not appear to be tied to any particular set of lock features or configurations, so unfortunately we cannot assume that one type of engraving is earlier than the other or that it necessarily comes from a different source.

Also pointing to the Pattern 1777 musket as the model for the Liège Besses is the flared-mouth second rammer pipe found on all of these muskets that have been studied. As was discussed earlier in the Pattern 1777 chapter, despite what might be indicated by incomplete Ordnance Department records, the Pattern 1777 musket appears to have had a flared second pipe from the very beginning. Indeed, the fact that a Pattern 1777 musket with a flared second pipe appears to have served as the model for Liège manufacturers at this early date only serves to reenforce that notion. Any gun that was supplied as a complete and finished weapon must be dated by its latest feature, not its earliest.

As an interesting point of comparison, we can look at the cost of a British-made Short Land musket and the cost of a Liège-made version during the years 1778 to 1780. Although it is vastly difficult to precisely cost out a musket produced under the Ordnance system, we can substitute those complete muskets as supplied directly to the Tower by Hirst and Pratt:

Supplier	Date	Cost in shillings (s) and pence (d)
James Hirst	1778–1780	32s 6d (or £1.12.6)
John Pratt	1778–1780	32s 6d (or £1.12.6)
Liège (w/bayonets)	Nov. 1778–July 1779	30s (or £1.10.0)
Liège (w/bayonets)	Sept. 1779–Feb. 1783	27s 6d (or £1.7.6)

Thus, we can see that up until mid-1779, the Ordnance was paying 2s 6d more (in coin terms, this amount was equal to a Half Crown, about the size of a half dollar and containing about half an ounce of silver) for British-made muskets, most often supplied without bayonets! After that time, the price of the Liège muskets and bayonets became even more appealing, with an increased savings of a whopping 5 shillings – the equivalent of one silver dollar per stand of arms! Multiply that by the 110,000 stands ordered, and the amount of money saved is tremendous.

Now it's time for the 9th inning cliché. You get what you pay for, and the Ordnance certainly did. Regardless of a savings ranging from about 8% to 16%, the Liège muskets and bayonets weren't up to good British snuff – a fact born out by the surviving arms themselves. Commonly found in American contexts, most specimens studied have either refaced or replaced steels, suggesting that this essential part was poorly made or that it wore out with unusual speed. Replaced cocks are also extremely common, indicating that the originals often broke. Since no specimens have been observed bearing British regimental markings of any sort, it can be assumed that they were deemed unacceptable for service by the regular regiments serving in the American theater.

Thus, it would seem that the majority of these muskets that were issued went to the various sorts of Provincial units and local Loyalist militias (those Americans remaining faithful to the King). Possibly bearing this out is a lone Liège Bess bayonet, found c.1900 during house construction in **Huntington, NY**. Since this Long Island town was more or less a Loyalist fortress during the latter half of the American Revolution, it was likely lost or discarded by an American in the King's service.

Afterwards, some of these "Arms bought in Holland during the War" were issued in new condition to the 44th Regiment in mid-1787, but they were described as "not good." It is no surprise that the Tower quickly rid themselves of these weapons, auctioning the majority of them in the mid-1780s, and sending others to remote garrisons.

Given the almost random variety of lock features that appear on these Liège Short Land Pattern muskets, the only thing that is entirely clear is that we don't have a very clear picture of what was going on. Perhaps a comprehensive study of the physical features on these muskets will eventually reveal an interesting story, allowing us to assign specific musket traits to specific contracts and deliveries. But as for now, it is all admittedly a bit of a muddle.

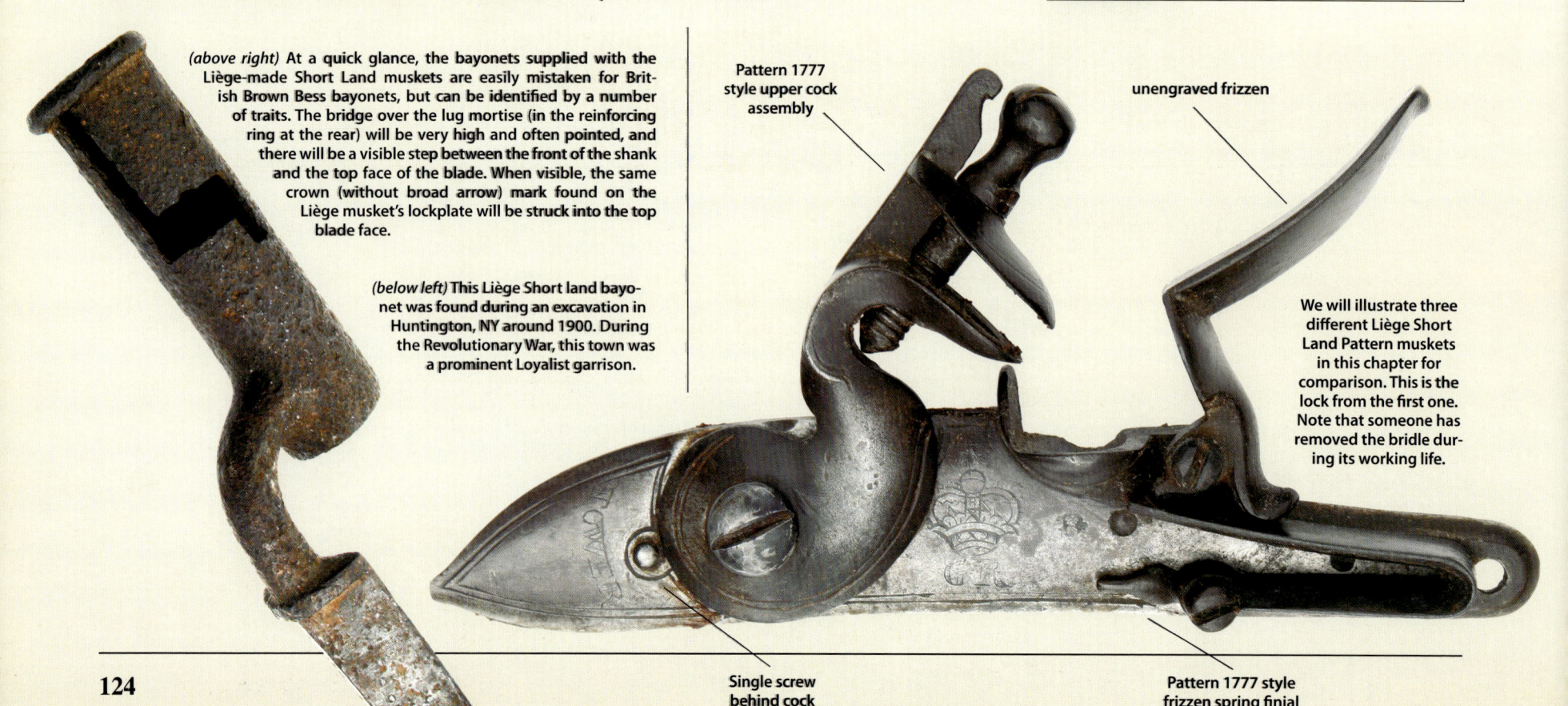

(above right) At a quick glance, the bayonets supplied with the Liège-made Short Land muskets are easily mistaken for British Brown Bess bayonets, but can be identified by a number of traits. The bridge over the lug mortise (in the reinforcing ring at the rear) will be very high and often pointed, and there will be a visible step between the front of the shank and the top face of the blade. When visible, the same crown (without broad arrow) mark found on the Liège musket's lockplate will be struck into the top blade face.

(below left) This Liège Short land bayonet was found during an excavation in Huntington, NY around 1900. During the Revolutionary War, this town was a prominent Loyalist garrison.

Pattern 1777 style upper cock assembly

unengraved frizzen

We will illustrate three different Liège Short Land Pattern muskets in this chapter for comparison. This is the lock from the first one. Note that someone has removed the bridle during its working life.

Single screw behind cock

Pattern 1777 style frizzen spring finial

Liège Short Land Pattern

More details of our first example of a Liège musket. An instant identifier of these Liège pieces is the engraving of the lock marks. Compare the markings on this page with those on page 100.

On this particular example, note that the top jaw is engraved and the top jaw screw is **not** pierced.

(above) All known examples of the Liège Short Land Pattern musket have the British Short Land Pattern 1777-style flared second ramrod pipe as illustrated here. This is a key factor in identifying one of these muskets.

125

More views of our first example.

A Second Example...

Much like the Short Land muskets procured by Dublin Castle, the Liège muskets differ slightly from their British counterparts in almost all minor details. A side-by-side comparison of Liège, Irish and British musket components will bear this out.

Note that the lock screw heads are much more highly domed than you would expect on a British musket.

Compared to the British Short Land Pattern buttplate illustrated on page 98, this Liège buttplate seems unrefined. It is also much shorter and wider.

(far right) Note that the "B" on the wristplate covers a partially erased "G", showing that this gun changed Companies.

Liège Short Land Pattern

The illustrations of this second example of a Liège musket start on the right-hand side of the previous page. Compare the engraving style on this large lockplate crown to the one on page 125, which is very different.

On Liège musket lockplates, a small crown without a broad arrow is struck below the pan. British examples of this ownership mark always have broad arrows.

127

A Third Example…

This third example of a Liège musket has a replaced cock — a very common defect on these guns. In the course of researching this book, the authors were surprised how many Liège muskets have replaced, repaired or missing cocks and frizzens. Having so many parts break must have been quite bothersome, and can be taken as proof that the Liège muskets did not meet the high expectations of the British armed forces.

128

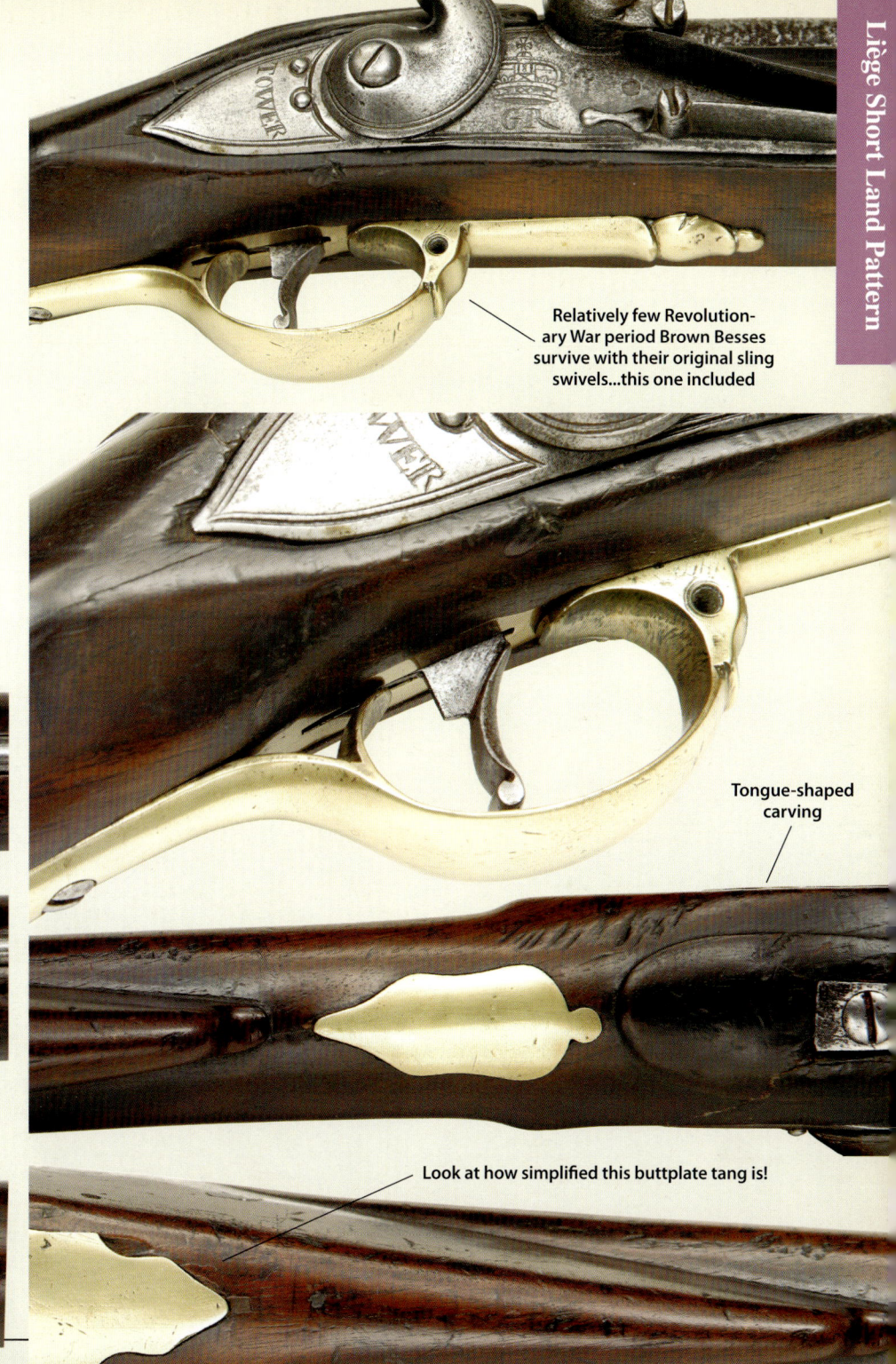

Liège Short Land Pattern

Relatively few Revolutionary War period Brown Besses survive with their original sling swivels...this one included

Tongue-shaped carving

Look at how simplified this buttplate tang is!

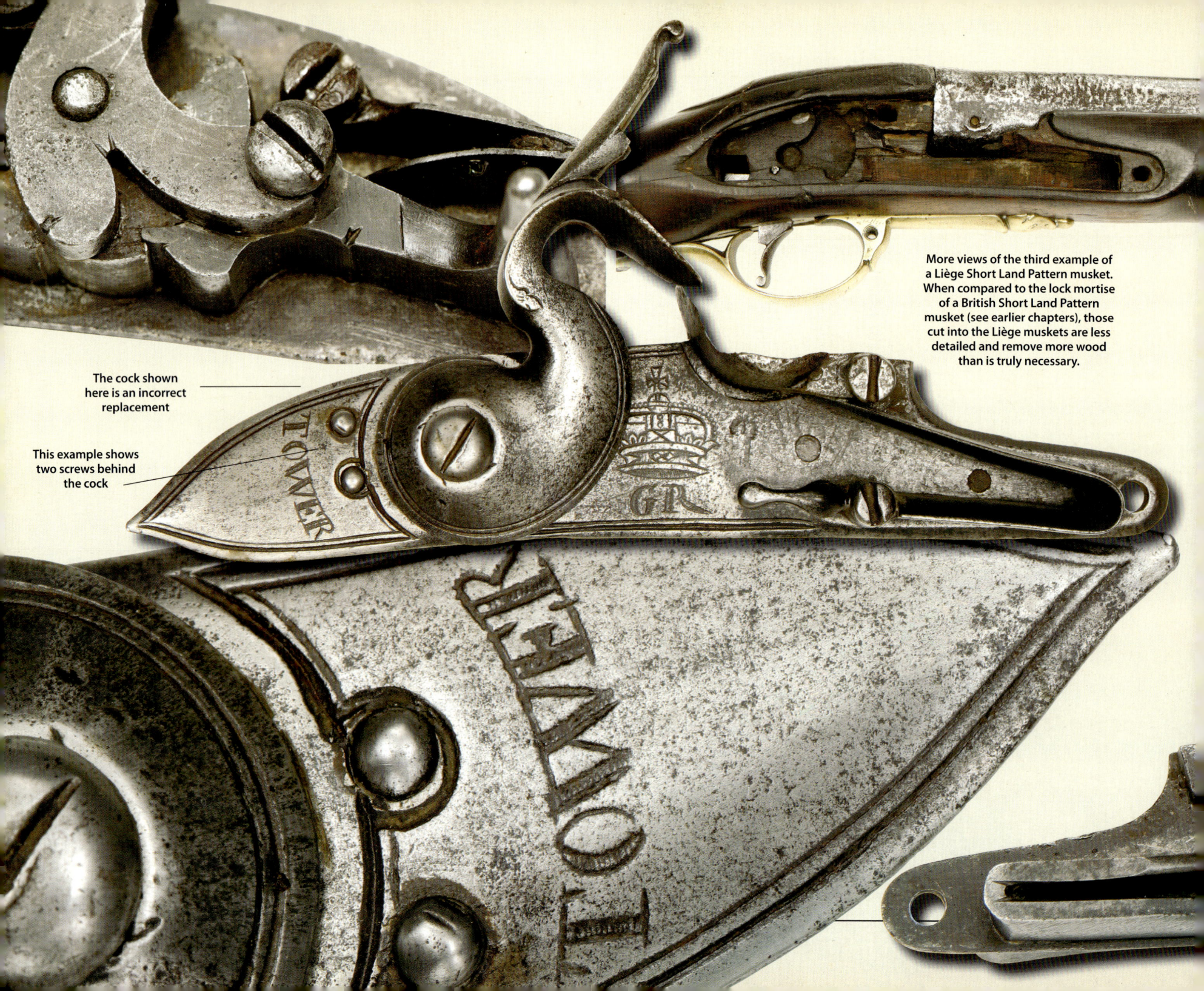

More views of the third example of a Liège Short Land Pattern musket. When compared to the lock mortise of a British Short Land Pattern musket (see earlier chapters), those cut into the Liège muskets are less detailed and remove more wood than is truly necessary.

The cock shown here is an incorrect replacement

This example shows two screws behind the cock

The lockplate on this musket has been engraved with a single-arched crown — something you would never expect to see on a British or Irish musket. See page 125 to view a Liège lockplate engraved with a double-arched crown.

Liège Short Land Pattern

(below)
Note the short sear spring.

131

THE PATTERN 1779-S SHORT LAND MUSKET

John Pratt supplies complete muskets with S-shaped sideplates.

Rarity: Rare (early type)

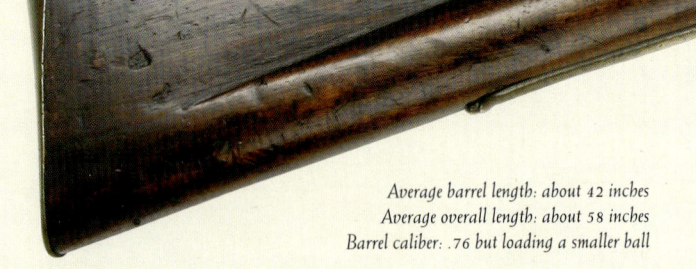

Average barrel length: about 42 inches
Average overall length: about 58 inches
Barrel caliber: .76 but loading a smaller ball

Long misunderstood as some weird, hybrid Short Land/Indian Pattern musket, this firearm is easily spotted by the presence of a convex, "S"-shaped sideplate on what would otherwise be a fully conventional Pattern 1777 musket. This unusual musket pattern was first brought to our attention by Anthony Darling in his groundbreaking 1970 book *Red Coat and Brown Bess*. Darling thought it was a 1790s weapon, pure and simple – and that certainly seemed logical enough. However, as it turns out, this musket was designed much earlier, during the American Revolution. It has now been established that these musket were made by John Pratt and collectors call them the "Pattern 1779-S".

In the Pattern 1777 Short Land Pattern Musket chapter, we discussed how wartime pressures forced the Board of Ordnance to abandon the Ordnance System in some cases, allowing British makers to supply almost 20,000 muskets in fully finished condition rather than as parts to be assembled. Well, this funny musket with an "S"-shaped sideplate is from that group. And as it turns out, of those muskets supplied as complete weapons between 1778 and 1780, only the Pattern 1779-S made by John Pratt has been positively identified.

As would be expected of a stop-gap measure, this arm fails to meet the quality of its Ordnance contemporary (the Pattern 1777). And it was definitely treated as an expediency, seeing almost immediate shipment for use by British and perhaps Loyalist units serving in the rebellious colonies. With such a small number of these arms being supplied (5,103 Pattern 1779-S muskets as compared to some 172,000 Pattern 1777s), the fact that numerous examples turn up in North America today in well-worn condition supports the idea that the Pattern 1779-S saw extensive use during the Revolution.

To date, two specimens bearing full British regimental markings have been identified: one marked to the 23rd Regiment and the other marked to the 76th Regiment (a.k.a. "MacDonald's Highlanders," disbanded in 1784). Both are marked in a different style, and likely received their engraved regimental designations "in the field," or perhaps at the anonymous American storehouse that issued them.

In addition to the arms issued to British line regiments from Ordnance stores at Dublin Castle or the Tower of London, tens of thousands of arms were issued from Ordnance stores in the colonies, most importantly those in New York City (one near the present site of the United Nations Building), Charleston and Providence. De Witt Bailey details these essential arms returns in Appendix VIII (pp. 320–329) of his excellent book, *Small Arms of the British Forces in America, 1664–1815*. In addition to being absolutely fascinating, these pages should be studied by every Brown Bess aficionado in order to understand the immense importance of these American arsenals and the sheer volume of their output during the Revolutionary War. Lost amongst these far-from-complete tallies is an accounting of the disbursement of a substantial portion of these wartime Pattern 1779-S muskets.

But wait...there's a catch! And it's a really big one! Just because a Pattern 1779-S musket is sitting in your hands, do not mindlessly assume you are groping a Revolutionary War musket. In correspondence with Bailey, he has made it known that he suspects more of these muskets were procured during the 1793–95 period. He just hasn't found the documentary smoking gun yet. However, it seems that some of the surviving Pattern 1779-S muskets are

Production Period:
First Period: April 1779 – July 1780
Second Period: post-1790 (possibly 1793–95)

Total Production:
First Period: 5,103
Second Period: unknown

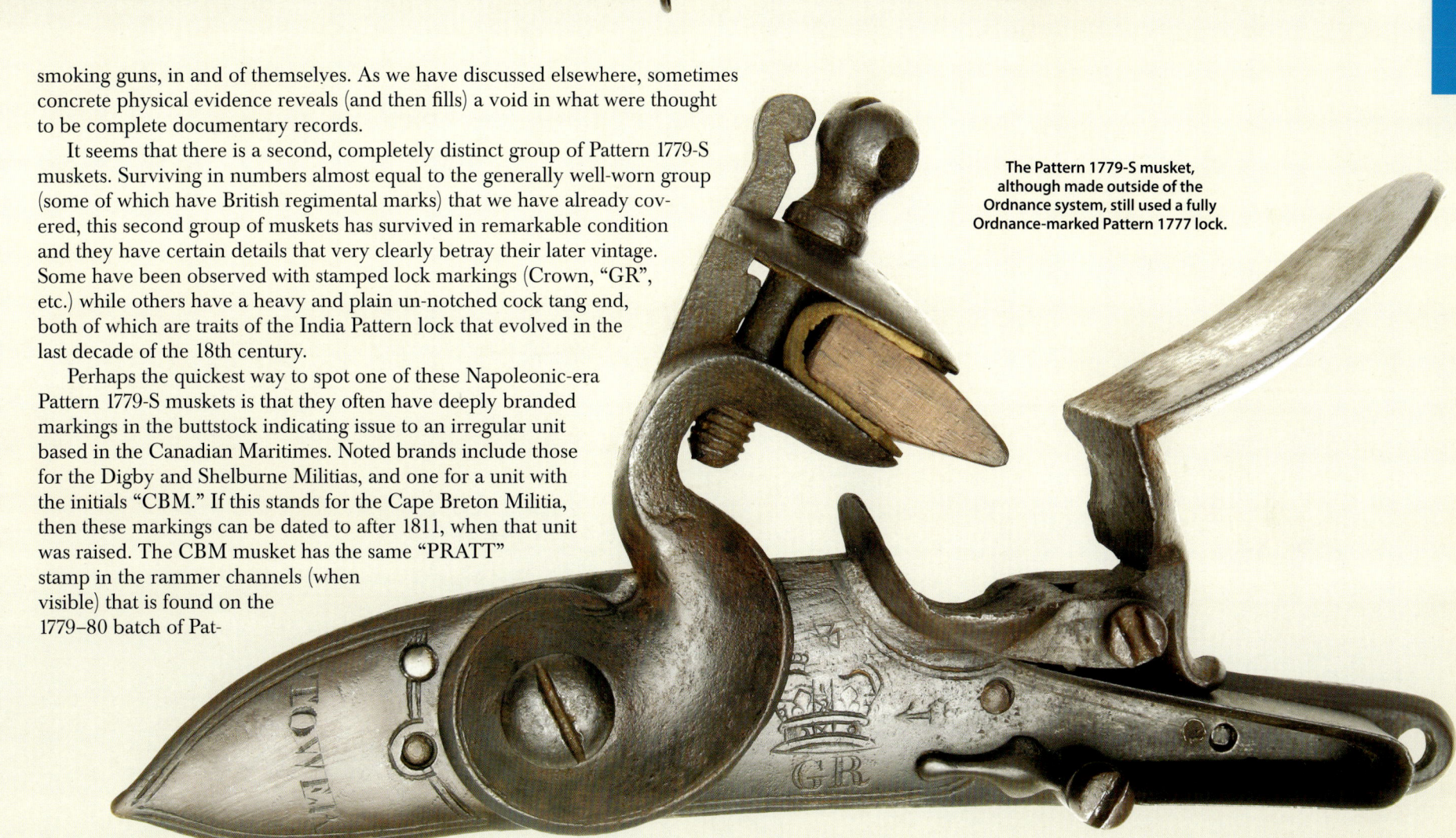

smoking guns, in and of themselves. As we have discussed elsewhere, sometimes concrete physical evidence reveals (and then fills) a void in what were thought to be complete documentary records.

It seems that there is a second, completely distinct group of Pattern 1779-S muskets. Surviving in numbers almost equal to the generally well-worn group (some of which have British regimental marks) that we have already covered, this second group of muskets has survived in remarkable condition and they have certain details that very clearly betray their later vintage. Some have been observed with stamped lock markings (Crown, "GR", etc.) while others have a heavy and plain un-notched cock tang end, both of which are traits of the India Pattern lock that evolved in the last decade of the 18th century.

Perhaps the quickest way to spot one of these Napoleonic-era Pattern 1779-S muskets is that they often have deeply branded markings in the buttstock indicating issue to an irregular unit based in the Canadian Maritimes. Noted brands include those for the Digby and Shelburne Militias, and one for a unit with the initials "CBM." If this stands for the Cape Breton Militia, then these markings can be dated to after 1811, when that unit was raised. The CBM musket has the same "PRATT" stamp in the rammer channels (when visible) that is found on the 1779–80 batch of Pat-

The Pattern 1779-S musket, although made outside of the Ordnance system, still used a fully Ordnance-marked Pattern 1777 lock.

Pattern 1779-S Short Land Musket

133

tern 1779-S muskets. This shows that John Pratt was the supplier of at least some of this later group, too. So, when confronted with one of these little beauties, look her over carefully! ❧

These two views show very clearly the distinctive S-shaped sideplate, which identifies this pattern of musket and helps to give it its name.

Pattern 1779-S Short Land Musket

The "SG" mark on the barrel stands for Samuel Galton.

Many British Land Pattern muskets have been struck with initials or letters on the sideplate flat — the meaning of these is unknown.

Since all of the Pattern 1779-S muskets produced during the Revolutionary War were supplied by John Pratt, his mark should appear in the ramrod channel. However, on well-worn specimens, it isn't necessarily legible (see below).

135

The green crud on the back of this sideplate is known as verdigris (Latin for "green grease"). This form of corrosion occurs when brass, which is a copper alloy, comes into contact with materials like wood and leather.

This Pattern 1779-S musket was unit marked in America following the Ordnance Style; this is musket number 23 in the 4th Company. For comparison, see page 118 in this book for an example of a wristplate engraved by the Tower.

136

(left)
A crowned 6 Ordnance inspector's mark struck into the stock just below the triggerguard.

From these angles, this Pattern 1779-S musket looks exactly like an Ordnance System-produced Pattern 1777 musket — showing just how carefully Pratt followed the pattern. This should not be surprising considering that Pratt was an Ordnance contractor building his muskets using many parts purchased from other Ordnance contractors.

Pattern 1779-S Short Land Musket

137

The initials "WBW" carved into the buttstock of this musket were added by an American owner.

Pattern 1779-S Short Land Musket

As can be seen from the proof marks on this barrel's breech, all of the Ordnance markings on this musket are heavily worn. Note the 23rd Regiment marking applied ACROSS the barrel.

(right) This "B" stamp near the toe of the buttstock was applied by a 20th-century collector to mark the musket as his. This is no longer considered proper.

By this point in the American Revolution (c.1779–80), the quality of engraving on Ordnance locks began to decline. There was a great amount of pressure to produce a large number of locks in a short amount of time and cosmetic perfection was no longer considered a top priority. See page 100 for a lock engraved just before the war.

Pattern 1779-S Short Land Musket

Details of the lock on this Pattern 1779-S. Even though Pratt supplied this arm complete to the Board of Ordnance, he used parts made by other contractors. This lock was made by Benjamin Willetts, and is identical to those that he was selling to the Tower under the Ordnance System during the same period.

The post of the cock on later-production Pattern 1777 locks are thicker and less graceful than this wartime product.

Note the wear to the frizzen spring caused by the forward movement of the frizzen when the musket was fired.

Note: The mainspring on this lock is a replacement, which appears to have been installed during the working life of the musket.

141

THE PATTERN 1793 MUSKET
(INDIA PATTERN TYPE 1)

Ordnance copies an East India Company musket pattern.

Rarity: Common

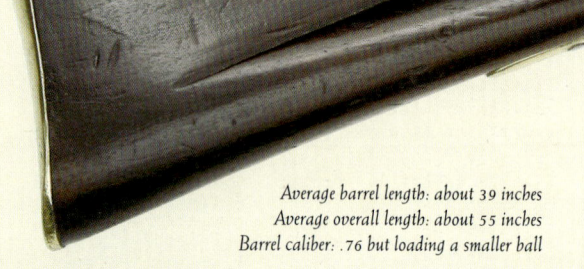

Average barrel length: about 39 inches
Average overall length: about 55 inches
Barrel caliber: .76 but loading a smaller ball

The execution of French King Louis XVI in January of 1793 turned the tide of British royal and governmental opinion against the French Revolution and drew Britain into a European fight once again. War with France was declared on February 1st, setting off a complex series of conflicts called the French Revolutionary Wars, which would stretch on until 1802.

Great Britain, however, was not prepared for war. In an effort to cut costs, their regular army manpower had been reduced severely, leaving an authorized strength of just 44,432 (including the colonies, India and the Irish establishment). However, not all of these slots were filled and the true effective strength was only about 36,500.

A rapid expansion of the forces began and this sudden demand for muskets was more than the Board of Ordnance could handle by themselves. Not only were they undermanned, but they were also unable to order very many muskets from outside suppliers because arms makers everywhere were busy with profitable contracts and the government had a reputation for being a difficult, picky customer who almost never paid on time. In October of 1783, the government decided to ask the East India Company (which had an extensive military of their own) whether they had any muskets available that could be diverted for wartime use. The Company had tens of thousands of muskets on order from gunmakers in Britain as well as completed weapons in storage at their warehouses awaiting shipment to India. These muskets weren't up to the normal standards of the Ordnance and their pattern was a bit different – but they were sturdy weapons and close enough to make do.

Beginning in very early 1794, a large supply of muskets was sent to the Ordnance from Company stores. They even took the extreme step of diverting shipments of arms already on their way to India. Later, when newly made muskets arrived from the Company's gunmakers, they were all sold directly to the government. By November 7th, a total of 28,920 muskets had been handed over. At that time, more than half of these muskets had already been issued to troops by the Tower. These transactions continued throughout the length of the conflict.

Sometimes, in order to speed delivery on Company musket contracts destined for government service, the Ordnance supplied barrels and locks to be made up into these "India Pattern" weapons. Under increasing pressure, the Ordnance itself began to have muskets of the India Pattern assembled and would buy any of these muskets that became available on the open market. The Board was in a desperate scramble for serviceable muskets of any description. They even allowed loose Short Land Pattern parts to be used in assembling muskets of the general India Pattern configuration. Short Land Pattern brass furniture, in particular, seems to have been used on these mixed-parts transitional pieces. Collectors should be aware, though, that there were also commercial muskets being made with mixed parts, which today are considered less desirable than their Ordnance counterparts. There are also other mixed parts muskets out there that have been assembled in modern times for fraudulent purposes. In order to qualify as a true Ordnance weapon, a musket of this period must have a crowned broad arrow on the lockplate, the King's view and proof marks on the barrel and an Ordnance storekeeper's stamp on one side of the buttstock.

By 1795, the Short Land Pattern was rapidly being phased out as the Ordnance was increasing

Production Period:
c.1794–1809

Total Production:
465,701 from 1795–1801, more than 293,810 after that

Lock Dates Observed:
Locks are not dated

Pattern 1793 Musket (India Pattern Type 1)

An Ordnance Pattern 1793 "India Pattern" Musket Marked to the 10th Battalion Royal Artillery

Note the "289" weapon number engraved on the buttplate tang of this musket.

their orders for India Pattern muskets. In 1797, the paperwork finally caught up with the reality of the situation and it was officially decided to standardize all future musket production (whether Ordnance or East India Company) upon the India Pattern until less desperate times when Ordnance might then be able to bring back the higher-quality Short Land Pattern. However, this never happened and the India Pattern became firmly entrenched as the new standard longarm of the British Infantry.

While the Ordnance Department was willing to purchase complete muskets in order to address shortages during the wars with France, it should be noted that the regular old Ordnance System of assembling muskets from separately contracted parts continued to chug along in the background. Interesting evidence of this comes from a February 1800 trial at London's "Old Bailey" criminal court in London. Robert Wright, who was a Board of Ordnance contractor engaged in rough stocking and setting up India Pattern muskets, was robbed of thirty musket locks that were the property of the King.

Wright's son, Michael-Memory Wright, testified:

I am in partnership with Robert Wright, No. 50, Prescott-street, Goodman's-fields. I am a gun-maker; we have a warehouse at the back of our house which communicates with the house, by a wall continuing on to the warehouse, which is about twenty yards from the house... I received information that the warehouse had been broke open... I missed a small parcel of musket locks, that were put by themselves, about thirty; they were worth about ten pounds... they had the government mark upon them...they are ordnance stores delivered to us from the Board of Ordnance.

Two robbers were apprehended fleeing through the

143

The locks built into Pattern 1793 muskets are functionally identical to the Pattern 1777 lock but differ in cosmetic details. Markings are now all stamped rather than engraved and all external parts are heavier and less graceful.

neighborhood with bags of musket locks under their arms and several locks dropping from the bags onto the ground. One of the robbers was a soldier in uniform. Robberies by soldiers were quite common during this period. However, most soldiers owned no clothing besides their uniforms, significantly simplifying their identification and arrest in those cases when there were witnesses.

The India Pattern musket that was at the heart of all this fuss was a common musket called the "Windus' Pattern" by the East India Company. The Windus' Pattern was purchased by the Company from 1771–1818. Compared to earlier Ordnance Land Pattern muskets, the Ordnance's India Pattern is quickly identified by its inelegant lock design, the relatively plain appearance of its brass furniture, and by the fact that its short, 39-inch barrel required just three ramrod pipes as opposed to the Short Land Pattern's four.

Overall, the India Pattern is not as well made as earlier patterns. It was subject to less rigorous view/proof, took advantage of quite a few manufacturing shortcuts and had stocks made from cheaper wood. On the plus side, though, it could be assembled more quickly and was a money saver for all concerned. In short, the Ordnance's legendary high standards gave way to urgent necessity and budgetary concerns.

Because of this compromise, because the pattern was borrowed rather than being an Ordnance creation, and because the Ordnance System itself was breaking down at this point in history, many collectors do not consider the India Pattern a true "Brown Bess." However, since the India Pattern is probably the most acquirable British flintlock military longarm out there, it would definitely be a disservice to our readers not to include it here, complete with a full set of detailed photographs.

144

Regardless of the many differences between the Land Pattern series of muskets and the India Patterns, there are still many similarities in form, as can be seen on this page.

This India Pattern buttstock profile is strikingly similar to those found on Land Pattern muskets going back to mid-century.

Pattern 1793 Musket (India Pattern Type 1)

145

The India Pattern introduced a completely different style of triggerguard.

With the India Pattern musket came a significant departure from the ramrod set-up of the Land Pattern series. The shorter barrel of the India Pattern required a rammer held by only three pipes. The style of the second pipe also became a shorter version of the trumpet style seen on the first pipe. Note that the upper sling swivel is now centered on the long trumpet pipe.

Note the simplified ends of the India Pattern's triggerguard.

The entry pipe remains essentially unchanged from Land Pattern muskets.

146

The India Pattern sideplate is chunky and lacks grace when compared to similar sideplates found on the Pattern 1779-S and earlier East India Company muskets.

Pattern 1793 Musket (India Pattern Type 1)

"D. Barrow" has yet to be identified.

147

The space between the flared end of the trumpet ramrod pipe and the nosecap is longer on India Patterns than on any Land Pattern musket.

(above) Crowned 14 Ordnance inspector's mark. Like Brown Besses themselves, the markings that the Ordnance applied to them evolved in form. A study of the other crowned numeral inspectors' marks that are found throughout this work will show a clear change in style over time.

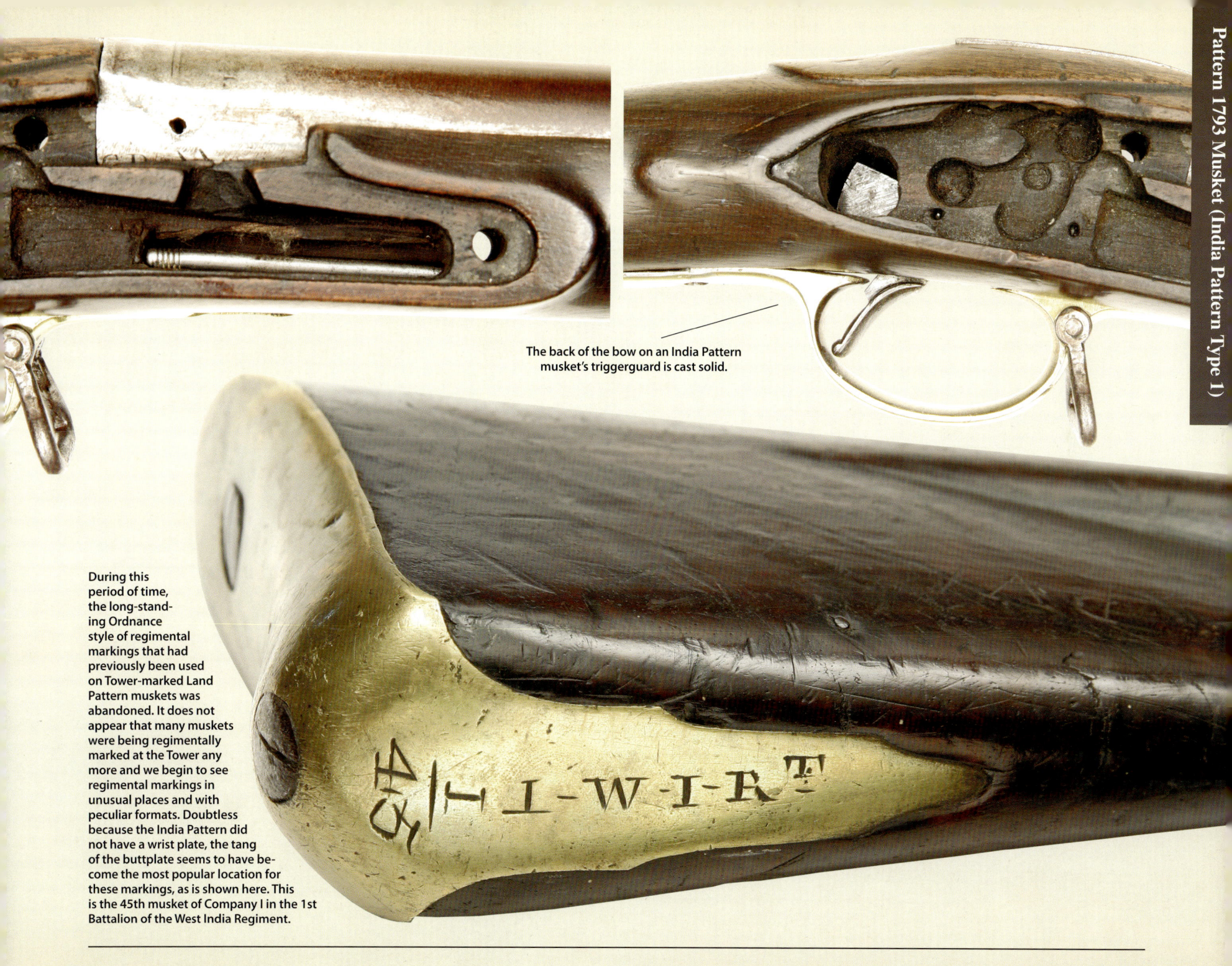

Pattern 1793 Musket (India Pattern Type 1)

The back of the bow on an India Pattern musket's triggerguard is cast solid.

During this period of time, the long-standing Ordnance style of regimental markings that had previously been used on Tower-marked Land Pattern muskets was abandoned. It does not appear that many muskets were being regimentally marked at the Tower any more and we begin to see regimental markings in unusual places and with peculiar formats. Doubtless because the India Pattern did not have a wrist plate, the tang of the buttplate seems to have become the most popular location for these markings, as is shown here. This is the 45th musket of Company I in the 1st Battalion of the West India Regiment.

149

150

Pattern 1793 Musket (India Pattern Type 1)

These detailed photographs show the various stamped markings applied to the lockplate of a Pattern 1793 musket. The blue color evident on the bridle and sear is the result of heat treating.

151

THE PATTERN 1809 MUSKET
(INDIA PATTERN TYPE 2)

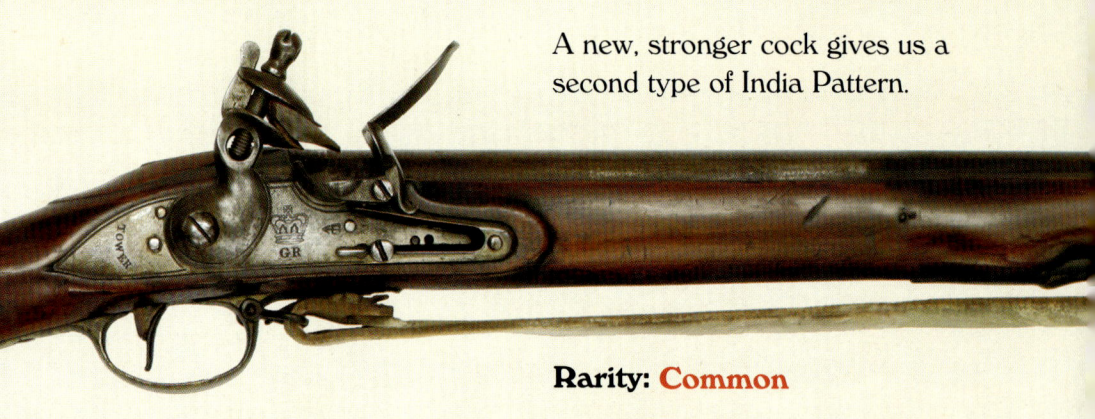

A new, stronger cock gives us a second type of India Pattern.

Average barrel length: about 39 inches
Average overall length: about 55 inches
Barrel caliber: .76 but loading a smaller ball

Rarity: Common

When peace with France was negotiated in 1802, the Ordnance saw this as an opportunity to replace the India Pattern musket with a newly designed weapon that they called the New Land musket. Unfortunately, peace lasted just one year and this sudden resumption of war scrapped all immediate plans to reequip the army with a new service musket. Orders were quickly switched back to the India Pattern and very few New Land muskets were ever produced or issued.

Although the New Land pattern was never officially abandoned, the India Pattern would remain the standard service arm of the British Army until the demise of the flintlock system. The only real change to the India Pattern design took place in 1809 and the resulting weapon is illustrated here. This new version is exactly the same as the preceding musket with the exception of two changes to the lock design. The cock has a more robust shape with a large circular opening at the center of its throat. This reenforcement made the cock less likely to snap in half, which was a weakness common to all flintlock designs. Less noticeably, the pan was also changed, providing a reshaped, deeper cavity for the priming powder.

Along with the earlier India Patterns already in service, the Pattern 1809 served the British infantryman during some of the most colorful episodes of his military history, including the Napoleonic Wars, the War of 1812 and countless colonial entanglements all over the globe. What had started out as a stopgap measure ended as a beloved symbol of British fighting spirit.

The total number of "Type 2" India Pattern muskets manufactured is unknown, but production was extensive, especially during war years. For example, 267,654 were set up in Birmingham during 1812 with 278,932 following the next year. In Birmingham alone, 1,341,625 India Patterns were produced between the years 1809–1815, when production was shifted to the New Land Pattern musket. This is in addition to those muskets being supplied by London makers like Ezekiel Baker, William Parker, Thomas Reynolds and Durs Egg. To give an idea of the extent of the London contribution, between the years 1803 and 1816, 845,477 muskets were manufactured in London, either by using the Ordnance System or as direct contracts for completed muskets. It is unclear how many of these muskets were "Type 2," but if overall production rates by year are anything to go by, it would seem to be considerably more than half of the total. A further hint is given by the fact that 2,085,974 India Pattern barrels were produced for the Board of Ordnance in Birmingham between the years 1809 and 1815, with 2,077,034 India Pattern locks being made for the Ordnance during the same period. So we are certainly talking about a lot of muskets here, and this pattern is by far and away the most common British flintlock military musket encountered today.

The power of Birmingham as a gunmaking center is easy to see in the figures quoted above. The trade had so thoroughly shifted from London to the north by 1798 that the Ordnance had to purchase land in Birmingham and erect viewing rooms. Also remember that the Ordnance muskets were only part of the story. Birmingham smiths were also cranking out weapons for the East India Company and the civilian trade at the same time. Almost five million longarms were produced in Birmingham during the war years 1804–1815.

When Napoleon was defeated at Waterloo in June of 1815, peace was restored to Europe and production of muskets slammed to a halt. In July, the Ordnance cancelled all outstanding orders

Production Period:
1809–September 1815 with a few being made sporadically at later dates

Total Production:
at the very minimum 1,341,625 but more likely about 2 million

Lock Dates Observed:
Locks are not dated

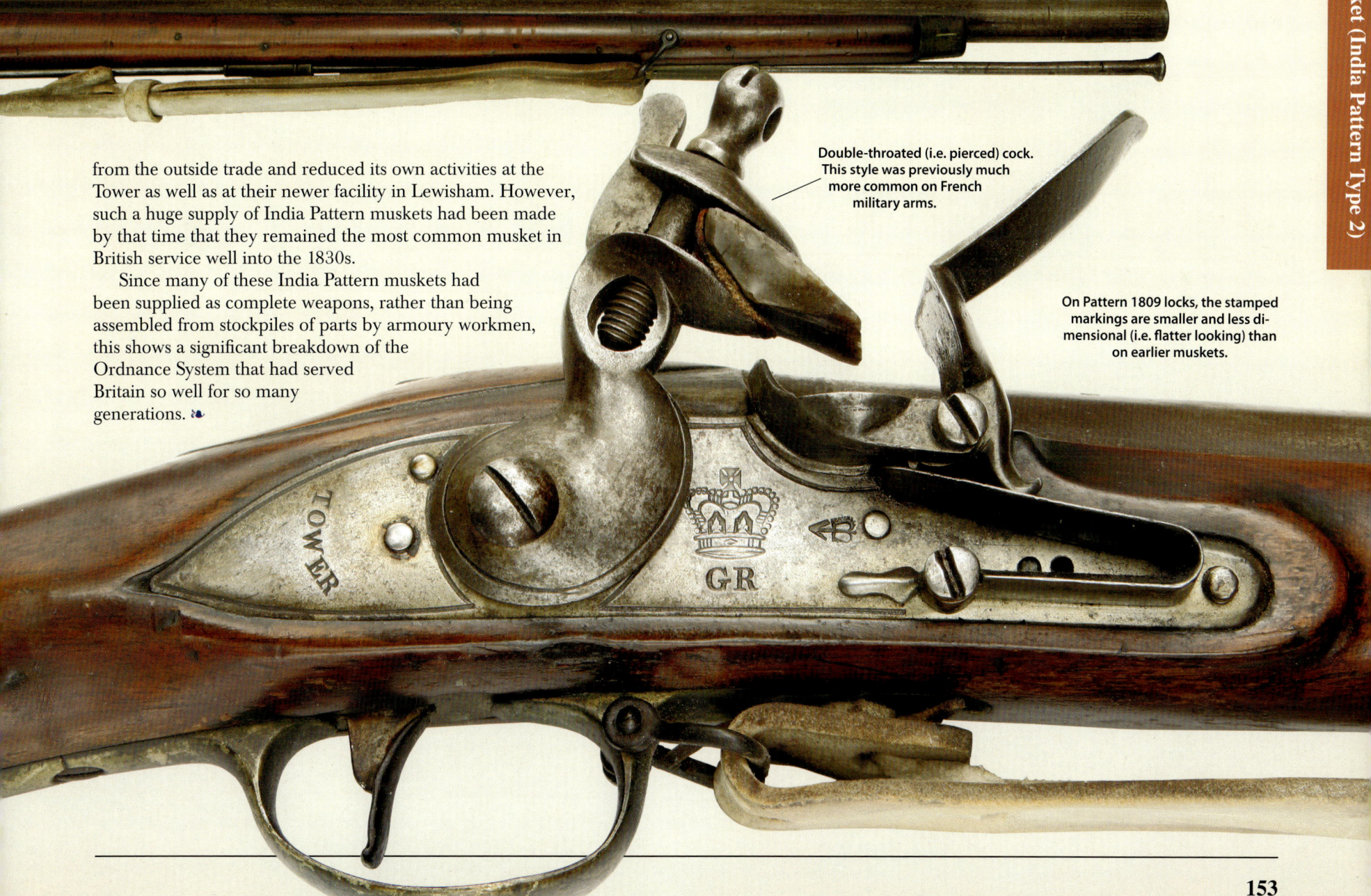

Pattern 1809 Musket (India Pattern Type 2)

from the outside trade and reduced its own activities at the Tower as well as at their newer facility in Lewisham. However, such a huge supply of India Pattern muskets had been made by that time that they remained the most common musket in British service well into the 1830s.

Since many of these India Pattern muskets had been supplied as complete weapons, rather than being assembled from stockpiles of parts by armoury workmen, this shows a significant breakdown of the Ordnance System that had served Britain so well for so many generations.

Double-throated (i.e. pierced) cock. This style was previously much more common on French military arms.

On Pattern 1809 locks, the stamped markings are smaller and less dimensional (i.e. flatter looking) than on earlier muskets.

153

The crude, engraved markings on this buttplate indicate issue to the first man (possibly an NCO) of Company H in the Royal Fuziliers, a.k.a. the 7th Regiment of Foot.

Like the Pattern 1793, the Pattern 1809 had a short, trumpet-mouthed second ramrod pipe.

Pattern 1809 Musket (India Pattern Type 2)

"The English…dull clods that they were, stood in brutish lines, unmoved by fine military music and the world-shaking clamor of assault columns shouting 'Vive l'Empereur,' until those columns came within a hundred yards: then they blew their heads off… generals, Eagles and all."

From the memoirs of Baron Marcellin de Marbot, John W. Thomasson translation, 1935

Made of whitened buff leather, the sling on this musket is believed to be original to the weapon's period of use.

Not completely breaking with tradition, the ramrod of this musket also carries the Company and musket designation "H" over "1" as found on the buttplate.

Pattern 1809 Musket (India Pattern Type 2)

This extreme close-up photograph of the King's cypher shows the simplistic nature of this marking when compared with earlier examples. This is due, to some great extent, to the fact that it is now stamped rather than engraved. This characteristic can also be observed on later British military firearms not covered in this book.

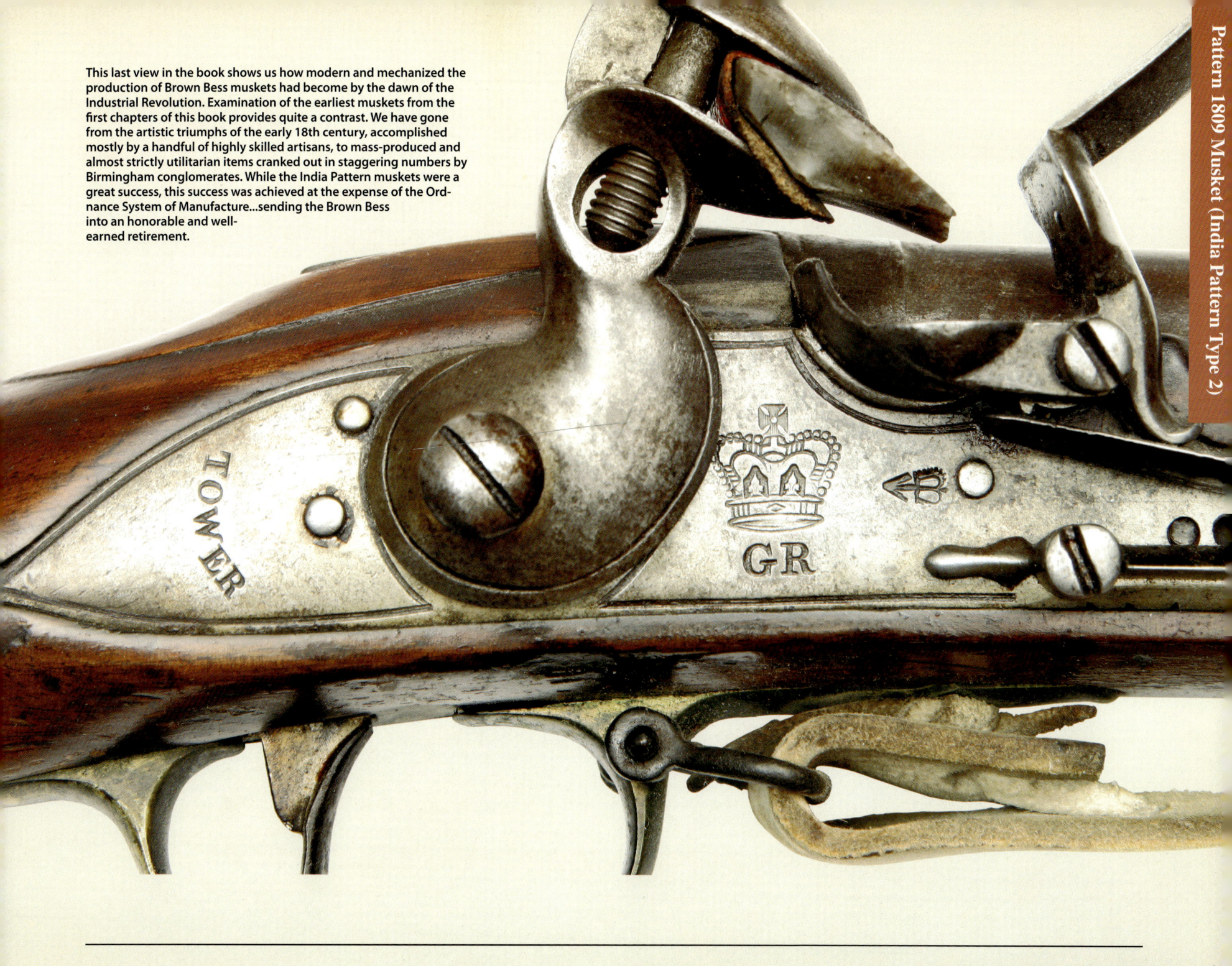

This last view in the book shows us how modern and mechanized the production of Brown Bess muskets had become by the dawn of the Industrial Revolution. Examination of the earliest muskets from the first chapters of this book provides quite a contrast. We have gone from the artistic triumphs of the early 18th century, accomplished mostly by a handful of highly skilled artisans, to mass-produced and almost strictly utilitarian items cranked out in staggering numbers by Birmingham conglomerates. While the India Pattern muskets were a great success, this success was achieved at the expense of the Ordnance System of Manufacture...sending the Brown Bess into an honorable and well-earned retirement.

Pattern 1809 Musket (India Pattern Type 2)

159

THE BIBLIOGRAPHY

AHEARN, William, "British Longarms at Lexington," *Man at Arms*, Vol. 20, No. 2 (April 1998): 15–24.

AHEARN, William, *Muskets of the Revolution and the French & Indian Wars*, Lincoln: Andrew Mowbray Incorporated, 2005.

ANONYMOUS, *The Succession of Colonels to All His Majesties Land Forces from their Rise to 1746*, London: J. Millan, 1746.

BAILEY, De Witt, *Pattern Dates for British Ordnance Small Arms, 1717–1783*, Gettysburg: Thomas Publications, 1997.

BAILEY, De Witt with VISSER, H.L., *Aspects of Dutch Gunmaking*, Zwolle: Waanders Publishers, 1997.

BAILEY, De Witt, *British Board of Ordnance Small Arms Contractors 1689–1840*, Rhyl: W.S. Curtis Publisher, Ltd., 1999.

BAILEY, De Witt, *British Military Longarms, 1715–1865*, London: Arms and Armour Press, 1986.

BAILEY, De Witt, "British Military Small Arms," *Journal of the American Society of Arms Collectors*, 71 (October 1994): 2–14.

BAILEY, De Witt, "Development of the British Ordnance Musket Lock, Part I," *International Collector Magazine*, Vol. 1, No. 1 (1995): 47–50.

BAILEY, De Witt, *Small Arms of the British Forces in America 1664–1815*, Woonsocket: Andrew Mowbray Incorporated, 2009.

BEDFORD, Clay P. and GRANCSAY, Stephen V., *Early Firearms of Great Britain and Ireland*, Greenwich: New York Graphic Society (for the Metropolitan Museum of Art), 1971.

BLACKMORE, Howard, *British Military Firearms, 1650–1850*, London: Herbert Jenkins, 1961.

BLACKMORE, Howard, *Gunmakers of London, 1350–1850*, York: George Shumway Publisher, 1986.

BRYCE, Douglas, *Weaponry from the Machault, An 18th-Century Frigate*, Ottawa: Parks Canada, 1984.

CARROLL, Donald, "The Brown Bess Musket Regimentally Marked," *Gun Report*, Vol. 35, No. 8 (January 1990): 36–42.

CARY, A.D.L. and McCANCE, Stouppe, *Regimental Records of the Royal Welch Fusiliers, Vol. I*, London: Royal United Services Institution, 1921.

COLE, Benjamin, *The Soldier's Pocket Companion, or the Manual Exercise of our British Foot*, London, 1746.

DARLING, Anthony D., *Redcoat and Brown Bess*, Bloomfield: Museum Restoration Service, 1970.

EGLY, Ted, "A Musket of Captain William Hickman, British 23rd Regiment – The Royal Welsh Fusiliers," *The Brigade Dispatch*, Vol. VII, No. 1 (January 1970): 5–6.

GALE, Ryan, "A Soldier-Like Way," *The Material Culture of the British Infantry 1751–1768*, Elk River: Track of the Wolf, Inc., 2007.

GOLDSTEIN, Erik, *The Socket Bayonet in the British Army, 1687–1783*, Lincoln: Andrew Mowbray, 2000.

GOLDSTEIN, Erik, "Infantry Hangers of the Royal Welch Fusiliers, 1742–1784," *Man at Arms*, Vol. 20, No. 6 (December 1998): 18–23.

GOLDSTEIN, Erik, *18th Century Weapons of the Royal Welsh Fuziliers from Flixton Hall*, Gettysburg: Thomas Publications, 2002.

GOLDSTEIN, Erik, "The Regimental 'Brown Bess' Bayonet, 1754–1783," *Man at Arms*, April 2005: 22–25, 32–35, 38–39.

GROSE, Francis, *Advice to the Officers of the British Army*, London: G. Kearsley, 1783

GUY, Alan J., *Economy and Discipline, Officership and Administration in the British Army, 1714–1763*, Manchester: University Press, 1985.

GUY, Alan J., *Colonel Samuel Bagshawe and the Army of George II, 1731–1762*, London: Army Records Society, 1990.

GUY, Alan J., "Minions of Fortune, The Regimental Agents of Early Georgian England, 1714–63," *Army Museum '85*, The National Army Museum, 1985: 31–42.

HAGIST, Donald and GOLDSTEIN, Erik, "Short Land Muskets for the British Light Infantry in America," *Man at Arms*, December 2009: 18–25.

HARDING, David F., *Smallarms of the East India Company, 1600–1856*, Vols. I & II, London: Foresight Books, 1997.

HOULDING, J.A., *Fit for Service, The Training of the British Army, 1715–1795*, Oxford: Clarendon Press, 1981.

LESLIE, N.B., *The Succession of Colonels of the British Army from 1660 to the Present Day*, London: Society for Army Historical Research (Special Publication #11), 1974.

MOLLER, George, *American Military Shoulder Arms, Vol. I.*, Niwot: University Press of Colorado, 1993.

MOWBRAY, Stuart C., "Brown Bess Locks: An Exciting New Discovery About Their Makers," *Man at Arms*, August 2009: 25–27, 43–44.

MULLINS, Jim, *Of Sorts For Provincials: American Weapons of the French and Indian War*, Elk River: Track of the Wolf, Inc., 2008.

NEAL, W. Keith and BACK, D.H.L., *Great British Gunmakers, 1540–1740*, Norwich: Historical Firearms, 1984.

NEUMANN, George C., *Battle Weapons of the American Revolution*, Texarkana: Scurlock Publishing, 1998.

NEUMANN, George C., *History of the Weapons of the American Revolution*, New York: Harper & Row, 1967.

NEUMANN, George C. and KRAVIC, Frank J., *The Collectors Illustrated Encyclopedia of the American Revolution*, Harrisburg: Stackpole Books, 1975.

PARGELLIS, Stanley M., *Military Affairs in North America, 1748–1765*, Hamden: Archon Books, 1969.

PRIEST, Graham, *Brown Bess Bayonets, 1720–1860*, Norwich: Tharston Press, 1986.

STRACHAN, Hew, *British Military Uniforms, 1768–1796*, London: Arms and Armour Press, 1975.

TROIANI, Don, *Military Buttons of the American Revolution*, Gettysburg: Thomas Publications, 2001.